PANSPERMIA: THE EXTRATERRESTRIAL HYPOTHESIS

PANSPERMIA: THE EXTRATERRESTRIAL HYPOTHESIS

Written by Chris Topaz
San Francisco, Ca.
TYN Research, San Francisco, Ca.

Panspermia: The Extraterrestrial Hypothesis

© 2013-2014 Chris Topaz, San Francisco, Ca.

All rights reserved. No part of this book may be reproduced or transmitted in any form or by any means without written permission from the author.

The views and opinions of the author are his alone and do not reflect the policy or opinions of any institutions of learning or media outlets giving interviews of the author.

The Style manual of the Associated Press was used during the editing for this book.

Fair Use Quotes are authorized. It is advisable to consult your region's fair Use Quote guidelines; however there may be academic penalties if the proper references are not stated.

The images may be protected by various agreements. Using an image places the user under agreements with the image rights holder not under the publisher's or author's agreements.

1st edition the ebook 2012

2nd edition color printing 2014

Panspermia: The Extraterrestrial Hypothesis

Chris Topaz

Table of Contents

1. Research Proposal: Panspermia   14
2. Panspermia: Cell and Microbiology Origins   24
3. Hypothesis   144
4. Conclusion   146

List of Illustrations

Figure 1. Fossil H. neanderthalensis Skull   24

Figure 2. Kepler Mission   35

Figure 3. Victim of Virus   54

Figure 4. Europa   59

Figure 5. Fungi   62

Figure 6. QE2   62

Figure 7. Mars   68

Figure 8. Mars   69

Figure 9. Day 679   71

Figure 10. Day 698   72

Figure 11. Mars Topo   73

Figure 12. Mars Fossil   75

Figure 13. Mars Fossil   76

Figure 14. Fungus   80

Figure 15. Mars Ocean   82

Figure 16. Yellowstone   85

Figure 17. E. coli   88

Figure 18. Arthropods   96

Figure 19. Crab   99

Figure 20. Fish Fossil   101

Figure 21. Fish Fossil   103

Figure 22. Acanthostega, or Panderichthys? 106

Figure 23. Anthropomorphic Amphibian   108

Figure 24. Primate   110

Figure 25. H. neanderthalensis   121

Figure 26. E. coli   149
Figure 27. Mars   151
Figure 28. Meteorite   152

Introduction

Life arrived on this planet from the Universe.

Contributors to this Book

To get from point A to point B you really have to arrive from a perspective that may be far from where you thought you were.

Chris Topaz

Chapter 1

PANSPERMIA
Abstract

Panspermia theory came up in the 19th and early 20th century as an idea of the possibility of existence of bioorganic molecules, seeds, germs, and organisms in space and how they reached the Earth. According to Arrhenius and Kelvin, Panspermia divided into two branches, which are lithopanspermia, defined as transportation of germs in stones that travel in space and radiopanspermia, which is the transportation of pores through pressure that radiates from stellar light. There has been the recent discovery that microbes can survive under extreme weather conditions which has been a credit to Panspermia hypothesis. There has been a laboratory test of cells falling to Earth at rates of a few tons in a day over the globe. The tests have proved positive by showing the existence of microorganisms in the cells, which are not far from the related species found on Earth. Bioastronomy is the field used to evaluate the existence of life in space and on other worlds.

Introduction

Charles Darwin in the 18th century accepted the pluralism of life within the cosmos. That was what helped him in his attempt to explain the origin of the variety terrestrial forms of life.

After decades of scientific investigations and continued questioning about the extraterrestrial life, a research on the universe and questioning of the possibility of universal life has developed markedly under astronomy. Panspermia theories established in the 19th century facilitated the evaluation space life, even though panspermia has no address to the first origin of life, but concentrates its argument on the continuation of life once an achievement of the origin comes.

There has been an expansion of microbial life limits in the Earth over the recent years with a wide variety of habitat. The habitats are found in the geothermal vents, in the ocean floor, in radioactive dumps and in Antarctic soil about eight kilometers beneath the Earth crust. Bacteria surviving for billions of years entangled in a crystal of salt are cases held with lots of weight in the field of astrobiology.

Panspermia Theory Explanation

The term panspermia derives from Greek words: 'pan' means all and 'sperma' means seed. Panspermia is an umbrella term that defines every scientific theory that suggests that all life known in earth had its beginning in the outer space. It assumes that survival exists in another place in the universe and terms this life as a catalyst to its survival on earth.

Scientists have defined the theory from two perspectives: The cosmic ancestry- in its view argues that life never began, but it existed all through in every part of the universe. In its opinion, there is no transportation of life to earth that occurred, and life got no origin. The earth formation in the Big Bang wake gave rise to a new planet where living microbes formed and took residence in it. This process was what repeated in the whole universe and innumerable places. The second view is that the earth at one time had no life hence the life ingredients originated from somewhere else in space.

Panspermia theorists have all along been debating on how life of these living microbes came to Earth from space. This debate has further categorized them into: Undirected Panspermia who have a long history in science argue that life ingredients came to Earth through no form of intelligence, extraterrestrial or divine nature stating that the whole process was basically random. Direct Panspermia advocates have assumptions that divine or even extraterrestrial intelligence aided as a catalyst to the seeding of life.

According to the majority of the scientists, Panspermia summarizes in the following ways.

a. Panspermia is about the extraterrestrial origin of essential building blocks of life that designates the start of the evolution process.

b. Panspermia is about the explanation of either how evolution was possible on Earth or elsewhere.

c. To embrace Panspermia, one has also to accept evolution. The theory deduces the evolution of seeded life from space but not mature life forms transportation through space.

d. If creation were as a divine deed that brought every form of matter to existence, then creation is compatible with Panspermia. It can show how life components and their distribution in the whole universe and on Earth took place in order to bring life.

Evaluation of Panspermia Hypothesis

Although some scientists argue the facts of the existence of intelligent extraterrestrial life as true, the scientific community have no such evidence thus far. Hence, discussions on Panspermia concentrate on the theory based on random distribution, presence of microbial life and its path to Earth from the Universe. This existence of life depends on answers to questions like:

1. Is there proof of existence of microbial life in space?

2. Is there proof that the life came to the Earth from space?

3. Is there a surety that primitive life in the Earth out of which advanced forms of life are said to have evolved were not in the Earth all along, and had never existed in space?

The scientists in defense for their theory give a response, as, on the issues of extraterrestrial life, scientists have been able to detect pre-biotic chemicals presumed present on the onset of evolution in comets, interstellar clouds, and meteorites. This gives them hope of that some raw chemical ingredients of life might have originated from space. These elements however do not meet the actual forms of microbial life.

The discovery of the meteorites in the Earth from Mars has proved to be the most promising answer to interstellar travels. These meteorites were first in three places namely, Chassingny, France in 1815, Shergotty, India in 1865, and Nahkla, Egypt in 1911.

The tiny tardigrades survival renewed confidence in Panspermia's theory. Tardigrades are speck-sized things with not more than 1.5 millimeters in length. They live on mosses and wet lichens and can wait for return of water once these environments dry out. They are able to resist cold, radiation, and heat. Tests in space that involved exposing them to solar radiations for ten days demonstrated the possibility of simple and tiny life forms surviving in space. This test on survival is of importance to Panspermia theory. Recent imagery from 'Mars Global Surveyor' and 'Mars Odyssey' revealed a number of branched valleys forming tightly packed and joined drainage systems showing an abundance of water on Mars at a certain time and that smaller melted amounts might still be on the surface of the planet. This is away from the consideration that the ultraviolent light presently attacking the place is an obstacle to the existence of organisms. This however, does not rule out the presence of life on Mars.

Human Genes Origin

Scientists have discovered that life in the Earth is by an assembly and maintenance by an information system which holds information contained within the living cell of every living organism. This information system is the DNA (Deoxyribonucleic Acid). Every living thing has a great amount of the information, contained in the DNA. The work of DNA is to preserve and to process information through highly complex function language. DNA contains the design needed in the construction of every part of all living organisms and the procedure of how to bring that design to life. DNA nucleotides arranged in a certain order along the DNA strands that form separate segments referred to as GENES. A gene is therefore, a specific segment of the DNA that contains the information to construct one specific protein for example, a gene to make hemoglobin, gene for myosin among other types of protein.

Origin of human from a DNA

A DNA molecule has 23 pairs of chromosomes the mother contributing half and the other half by the father. In this DNA, all information necessary to build, grow, protect, and maintain thinking human is present. The DNA in this zygote assembles all raw materials within the cell and goes ahead to use them in the construction of embryonic cells. It continues to differentiate cells further into lungs, eyes, kidney, stomach, and all the rest. After nine months, a baby is born who continues to grow up to a time when it is now a mature adult. During the growth to adulthood, speech develops hearing, and walking among many others.

Intelligence from Panspermia theory perspective

Some astronomers argued that intelligence does not occur elsewhere in the galaxy. They argued that only a computerized robot would exist here. One theory used to disprove this is by the research of Frances Crick the co- discoverer of DNA structure. He suggested that there was an advanced civilization, which might have spread simple bacteria and other biological constituents all over the cosmos arguing that the coming of these materials on Earth started the evolution chain, which led to a man. Perhaps as life could be universal, then so too is speech and perhaps even intelligence is universal.

References

Crick, F.H.C., Orgel, L.E. (1973). Directed Panspermia, Icarus 19, pp 341-346.

Dawkins, R. (1976). The Selfish Gene, Oxford University Press, Oxford.

Horneck, G., Bücker, H. and Reitz, G. (1994) Long-term survival of bacterial spores in space. Space Res, 14, 41-45.

Horneck, G., Eschweiler, U., Reitz, G., Wehner, J., Willimek, R. and Strauch, K. (1995) Biological responses to space: results of the experiment "Exobiological Unit" of ERA on EURECA I. Space Res, 16, 105-118.

Hoyle, F. Wickramasinghe, N.C. (2000). Astronomical Origins of Life, Steps towards Panspermia. Kluwer Academic Press.

Joseph, R. (2000). Astrobiology, the Origin of Life and the Death of Darwinism, University Press.

Barber, D. J., and E. R. D. Scott, "Origin of supposedly biogenic magnetite in the Martian meteorite Allan Hills 84001," Barber, D. J., and E. R. D. Scott, "Shock and thermal history of Martian meteorite Allan Hills 84001 from transmission electron microscopy", Meteoritic and Planetary Science, 41(4), 643-662, (2006)

Joseph R. (2009a). Life on Earth came from other planets. Journal of Cosmology, 1, 1-56.

Chapter 2

Figure 1. H. neanderthalensis. (Getty Images: iStockPhoto)

> *"Athena had breathed life into the clay figures of man created by Prometheus and Epimetheus who were spared imprisonment in Tartarus because they had not fought with their fellow Titans during the war with the Olympians. Prometheus had shaped man out of the mud of the primordial Earth. Thusly they were given the task of creating man whom the Gods did not particularly like at first…"*
>
> —Paraphrase of ancient history

Was the gene that made possible animal life from a Panspermia event and if so could the language gene(s) in H. sapiens be traced to a foreign gene?

Panspermia: Cell and Microbiology Origins

Even though all chemical reactions for life can be traced to conditions on primordial Earth, Panspermia can explain cosmo-genesis on planet Earth because organic molecules have been observed in space by telescopes and Mars appears to have conditions for life, and at least one life form observed by planetary robotized model rovers. If the case could be built for Panspermia, then what was the origin of the Human speech gene since it seems to pop in from somewhere else?

The Panspermia Theory

The British Astronomer Sir Fred Hoyle was famous for coining the term Big Bang and also believed that life was seeded in the Universe by microbes. In the case for the big bang he proposed that it couldn't have existed because the Universe did not have space or time yet in existence for anything to expand in to. If there was a big bang then where did the medium come from that is propagating the matter and energy in the first place? This has given rise to recent theories of the Multiverse. He was an advocate of Panspermia and rejected the abiogenesis model of the origin of life on Earth. According to Hoyle the chances for a cell to acquire the set of enzymes required for life would be $10^{40,000}$ while the number of molecules in the Universe is 10^{80} thereby proving in his calculations that something unusual went on in the primordial creation of the first cell on Earth (Hoyle, P. 35,19 81). The calculations and worked equations are the basis for the argument of Panspermia. He advocated that life must have come from a space seeded debris such as that offered by the observations of comets.

I have built computer programs in academic computer programming classes to calculate every elementary particle in the Universe. It is a common assignment. The assignment usually attempts to get the student to conserve machine operations and the offer is made that the mainframe is available if you need it. To navigate the assignment correctly usually a small machine level program is constructed or if it is made in an object oriented language then a function is made where some of it is chained, and some of it uses recursion. I used the dimensional analysis method out of chemistry and created a series of functions. It was an experiment to see how many different arguments can be made for the answer for the class. In any case the actual number of particles in the Universe is surprisingly agreed upon to such an extent that lower division college students routinely make the calculations for their assignments. No one questions the number of particles in the Universe; therefore it can be a basis for a hypothesis.

To Summarize:

Knowing the amount of particles in the Universe; there are not enough particles in the Universe for the probability to exist for enough collisions to form the enzymes needed in the first animal cell. There must have been at least two sources of starting materials to form the first animal cell.

In the preceding Hoyle's calculations: there must be other ways that genetic materials are blended with Earth's genetic materials because the probabilities are so low and the chances of combinations happening are far too remote.

Panspermia can be a theory used to explain what is described in the following.

The Universe is defined as the region of space that we can see with our telescopes in a sphere. Some include the region that is just outside of that region. So by definition the region outside of the said sphere is another Universe. That is not suggesting a parallel universe but a Universe beyond our detection of the same laws of physics that we currently operate under. Perhaps the sphere could expand in a given amount of time due to various occurrences and the Universe would get larger. My thoughts are that the big bang that we are detecting with all of the microwave background radiation is the previous super nova that caused the nebulae to form where our star formed. This brings the big bang to a more local area of the Universe. It also means that in fact what we are calling the Universe is in fact the local area of each star. Each star is its own Universe. The Universe, the multiverse is more accessible and yet more unattainable now with the perspective.

In our solar system we have an array of planets where all of the components of life as we know it exist. Every planet in our solar system has trace amounts of water. Various planets have organic compounds. In an intuitive sense that is all we need for life to form. Our solar system has all of the components of life.

Yet what we have are clear theories with distinct boundaries between them:

Life formed only on Earth

Life formed on Earth with organics from elsewhere.

Life formed elsewhere and arrived on Earth.

The truth can be determined by putting everything into a syllogistic reasoning equation and running the logic through. This logical analysis will be presented later.

The Drake Equation

The Drake equation indicates that intelligent life should be common. Somewhere a life form's animal pieces could have been incorporated into the life on our planet through asteroid impact(s) or other Panspermia events.

Using the Drake equation calculate the number of intelligent extraterrestrial civilizations with which we might be able communicate. The further investigations of the Drake equation must be conducted by Clinical Anthropologists.

The premise for much of this is flawed. Someone (Drake) is under the assumption that an ET civilization will somehow send out a homing beacon for us to pick up. The other assumption is that these ET civilizations will somehow use radio frequencies the same way we do. The radio beam theory states that somehow the carrier wave in an AM modulation system will somehow be used to either contact us directly or that the signal will leak in our direction. No credence is give that an ET civilization instead of sending passive aggressive beam communication that instead they might simply beam their ships around the galaxy or directly here. The SETI programs instead create somewhat of a circular argument that we chase our tails looking for their Citizens Band hailing frequency. The idea of a superluminal carrier wave is never considered by any one in these discussions. A superluminal wave is based upon a carrier wave, and a signal riding on the crests of the carrier wave thereby surpassing the speed of light. These superluminal techniques could either beam a signal or modulated matter in a direction. The idea that contact will begin in the future according to the Drake equation creates an assumption that there has been no contact already, or that radio communications are used in a certain way.

Drake equation assumptions based upon my research.

Logical Premises

The most common element in the universe is Hydrogen, and can be seen in the night sky in all of the billions of uncountable stars. It is element number one, on the periodic table of elements. This element forms water by simple oxidation. Water is found in every planet in our solar system, confirmed by many telescopic observations. It is known that water is necessary for life in the universe. With these facts life as we know it based upon water should be everywhere.

Organic molecules have been found in meteorites. Organic molecules have been observed on the planet Neptune, Mars and a moon of Saturn in our solar system. Organic molecules are very common in the Universe.

Clouds of Amino Acids have been confirmed in space by telescopic observation. Amino Acids are the building blocks of protein chains and the R.N.A. sequencing of proteins, as well as the basis of D.N.A. itself. These chemicals are made from the element Carbon, and are also very common in the Universe. Carbon based organic molecules, and Amino Acids are confirmed many times in space by astronomy.

Unknown fossils of life not from Earth were found, and cataloged, and taxonomy started based upon the data in the 1800's (Proctor, 1882). The fossils were the first confirmation of extraterrestrial life discovered. The individuals describing them didn't enter the primary literature however due to their educational status as not having any. The fossils were found in Meteorites and filed in University archives. The writings at the time caused quite a stir. The references and the arguments have been overlooked for quite some time. Never-the-less extraterrestrial life in the form of fossils has been cataloged since the 1800's.

Our civilization has already produced an accidental aircraft that left the solar system and may have been the basis for contacts to begin with E.T.s. A manhole cover was photographed going 67.56Km/ Second (after a nuclear test explosion), where the escape velocity of the Solar System is 35 Km/ Second. The manhole cover launched in 1957 left the solar system in 1961 going 2% the speed of light! Our history in this is actually sending flying disks like saucers going out of our own solar system.

These factors are what I considered when calculating the Drake equation in that I do not agree with this assessment in the first place. I do not know what form of the equation that I would use or that I would use this equation at all or what equation that I would come up with. However:

The Drake equation: $Nc = N^1 * Fp * Nlz * FL * Fl * Fs$

N^1 – Number of stars in the Galaxy.

Fp – Fraction of stars with planets (I agree about the number for optimistic). I think that even binary stars have lots of planets, so my actual number would be higher.

Nlz – Number of planets per star that lie in the life zone for longer than 4 billion years. Here I think that three planets at least have the ability to sustain life here in our solar system, Venus, Earth, and Mars. Mars has temperatures like that on Earth for a tall mountain and Venus has an atmosphere that may be teraformed.

Fl – Fraction of suitable planets on which life begins.

Fl – Fraction of planets where Intelligence forms.

FS – The most important factor is FS, the fraction of a star's lifetime during which a civilization is alive and transmitting. Our civilization as an example 100,000 years from the times of the unification of Egypt/ divided by the age of the Sun (10 billion years).

Nc = 2e11 * 0.5 * 3 * 3 * 3 * 1e-5

Nc = 27 Million Systems with intelligent life transmitting signals.

1. According to the Drake equation there should be 27 Million Extraterrestrial civilizations in our galaxy. This would correspond to amino acids found in space, and organic matter found in meteorites. Life should be under every rock in the galaxy.

2. Based on the fact that there were batteries found in ancient Egyptian excavations, our total civilization should be around 100,000 years old. The elite members of these societies have claimed to be in contact with gods from outer space a premise still maintained to this day by every religious order on Earth. Therefore: our technological civilization has lasted at least as long as the civilization of ancient Egypt to the present day.

3. I think that contact may have already begun with extraterrestrials based upon the results of the Drake equation, and the amount of religious texts that claim as much. Every religion on planet Earth from antiquity complains about interventions from the extraterrestrials. We have also already launched a flying disk out of our solar system a virtual red flare in the cosmos, a beacon telling of our technological prowess. The Drake equation would indicate that life is plentiful.

What is interesting about the Drake equation is the fact that so many intelligent life forms are predicted and it is exactly what we see in so many hominid species on Earth. That is the puzzle: the Drake equation itself seems to predict the amount of intelligent hominids found terrestrially due to convergent evolution but it would not be the case intuitively.

There are other issues involving the first animal cell that formed from single celled organisms; specifically the event that occurred when the mitochondria progenitor became incorporated into the first animal cell as a symbiont. That is the thing: it would have been fine if not for the slight complication that Earth life has organelles that are symbionts from precursor prokaryotic life forms. A complex event that may never be understood has occurred. As to how it occurred and that it has resulted in multicellular life as we know it is a topic of the question.

The Telescopic Observations

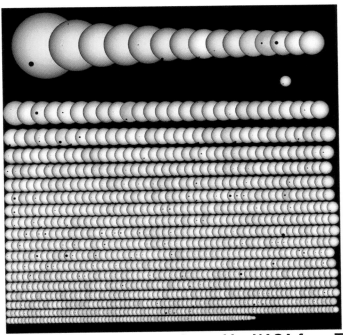

Figure 2. In this photo released by NASA from The Kepler Mission: Viable life planet candidates are lined up, Jan 07 2013. It is inevitable that life will be discovered soon on these or other worlds. (NASA)

A spacecraft was developed to specifically search for planets, habitable planets and planets in the Goldilock's zone of observable stars. It was called the Kepler mission.

> *"…The challenge now is to find terrestrial planets (i.e., those one half to twice the size of the Earth), especially those in the habitable zone of their stars where liquid water and possibly life might exist… (http://kepler.nasa.gov/)."*

The Kepler mission is of course extremely provocative if one has not been following advances in the sciences in regards to telescopic observations. The thought is that any time soon there will be found the first extraterrestrial civilization and that they will be close by.

In the cosmos what can be seen are remnants of building blocks of life from supernova explosions of stars in the surrounding gaseous debris. These are detected by placing instruments on the telescopes to detect how light is altered when arriving from the region of space. This would suggest that on the solar system's primordial basis there must have been the basis for life already present.

Organic Compound Found In the Stars

Life-building molecules might be spread throughout space.

Nature

Ewen Callaway

Published online: 24 July 2007

"Astronomers have found the largest negatively charged molecule so far seen in interstellar space. The discovery, of an organic compound, suggests that the chemical building blocks of life may be more common in the Universe than had been previously thought…"

"The molecule is a chain of eight carbons and a single hydrogen called the octatetraynyl anion (C_8H^-). Two teams of scientists have spotted it near a dying star and in a cloud of cold gas…"

"… the octatetraynyl anion in a dense cloud of gas in the halo surrounding a dying star in the constellation Leo, 550 light years from Earth. They made the discovery using the Green Bank Telescope in West Virginia, a radio dish 100 meters across…"

"Another team from the Harvard-Smithsonian Center for Astrophysics in Cambridge, Massachusetts, used the same telescope to spot the compound in the

Taurus Molecular Cloud (TMC), located 450 light years from Earth in the constellation Taurus. Both findings are published in the Astrophysical Journal…"

"…negatively charged organic molecules such as the octatetraynyl anion were long thought to be confined to Earth…"

"To show that octatetraynyl anion could be found in space…the two teams sifted through radio telescope measurements to find the molecule…"

"…The spectrum hinted at a compound called hexatriynyl anion, but confirmation only came when they made the compound and lab and found that the two signatures matched up…"

"The researchers believe that even larger organic molecules are likely to be out there…"

The materials can also be found around distant planets through telescopic observations.

Stuff of Life (but Not Life Itself) Is Detected on a Distant Planet

New York Times

By DENNIS OVERBYE

Published: March 20, 2008

> "Astronomers reported Wednesday that they had made the first detection of an organic molecule, methane, in the atmosphere of a planet outside our solar system and had confirmed the presence of water there, clearing the way for a bright future of inspecting the galaxy for livable planets, for the chemical stuff of life, or even for life itself..."

This result makes the thought that in a typical solar system all of the components of life are found all throughout the solar system and not just in the habitable zone itself.

Our solar system condensed from an even earlier solar system from at least one earlier type G star, which gave us the necessary components to condense our current star. What we came from can be determined from the surrounding sister stars and how much distance was traveled since the beginning of the solar system, which has been one quarter turn around the galaxy. We have rotated one quarter turn around and have traveled with our sister stars that came from the nebulae of gas that formed our star cluster.

The First Cell Membrane

Having a carbon yard full of organic debris floating around does not make life possible. Just the cell membrane itself is a complicated affair. The cell membrane would have to come along to gather in a bag all of the assorted organic molecules.

A cell membrane is often described as a phospholipid bilayer consisting of hydrophilic carbon heads facing outward and hydrophobic carbon tails facing inwards. The cell membrane likes water on the outside and oil on the inside of itself.

We know from the study of animal life on Earth that the types of phospholipids are dependent on where the animal lives, for example: the ice fish, etc. The descriptions of the phospholipids are that they have kinks in their tails dependent on the environmental selection, etc. The quality of the animal in regards to if we should eat it or not depends on the amount of kink to the phospholipids in their tails of the phospholipid bilayer of their cell membranes.

We know a lot of cell membranes yet there are no places that we know of where blank cell membranes float around in nature – they must be manufactured by mechanisms.

The Pre-biotic RNA World

There is a hypothesis among microbiologists that goes unstated: there was an RNA world everywhere out there or here at one point or even currently. It is inherent when undertaking the daunting task of becoming a microbiologist. The view is from what microbiologists know of the mechanisms of life. Using an RNA template manufacturing of proteins could be carried out by the same mechanisms as that of which is found in cells. In these theories an unstated hypothesis goes that: as a matter of discourse RNA functions so similar to DNA that of course molecules self-assemble and carry on basic processes just in a form of chaos going against the gradient of the entropy law of the thermodynamics of the Universe.

The smoking gun is of course the master enzyme: the ribosome. The ribosome takes an RNA template and manufactures proteins. The ribosome has changed very little across the eons from antiquity to the modern ages of life. The ribosome could be said to be the one surviving structure of all time proving a theory of organic-genesis.

The ribosome is fantastical in the shape of it. There are diagrams of the ribosome comprising of the individual nucleotides. The ribosome looks like it is one of the modern art pieces made with a 3 dimensional printer.

It is the function of the ribosome that really causes reflection. If it does not make you pause and contemplate it then you have not fully understood its' meaning.

In object oriented computer programming there is what is known as an object Iterator that steps through objects in a priority queue. When you have a list or stack of objects it is the Iterator that steps through and reads each object. The ribosome is what steps through a list of RNA and reads the code much like a programming Iterator.

That is what is fantastical about the RNA templates: the RNA templates are exactly like objects of code in object oriented programming where peptide synthesis is the result of the reading of the code.

A ribosome could be an artifact that would take everything and advance a process of life. If it were introduced somewhere with pools of organics around then it would start a chain reaction to start synthesis.

There would have been pools of organics lying about until someone or something tossed in a ribosome, and then all hell broke loose. It is putting out a fire with gasoline when you introduce ribosomes. Ribosomes are the ultimate enzyme.

Enzymes lower the activation energy of organic bio chemical reactions thusly taking entropy and standing it on its ear. What is required is the right enzymes to be present at the right times.

These theories of the pre biotic world of RNA synthesis are validated by the astronomers who observe clouds of nucleotides out in space formed in various processes out in the cosmos.
Everywhere that you look in a telescope in every direction the nucleotides are seen in space. There seem to be nucleotides out there in the Universe floating around naturally in space thereby reinforcing the notion of biologists that there existed a hypothetical RNA world of molecules based upon what is known of the processes of life conducting the chemical reactions present at all levels of cellular activity.

The DNA

DNA is so novel that It has been investigated to determine if it is older than life itself. The molecule is like an artifact of unknown and/ or complex authorship. The relationships that can be generated from the base codes are simply transcendent.

DNA could have existed long before life itself

New Scientist

24 August 2012

by Michael Marshall

"After decades of trying, in 2009 researchers finally managed to generate RNA using chemicals that probably existed on the early Earth."

"That could have important implications for our understanding of life's origins. Prebiotic chemists have so far largely ignored DNA, because its complexity suggests it cannot possibly form spontaneously."

"Conventional wisdom is that RNA-based life eventually switched to DNA because DNA is better at storing information. In other words, RNA organisms made the first DNA."

"…the story makes more sense if DNA nucleotides were naturally present in the environment. Organisms could have taken up and used them, later developing the tools to make their own DNA once it became clear how advantageous the molecule was - and once natural supplies began to run low."

Investigating DNA is like being an Archeologist of the Universe, attempting to piece together unknowable puzzles of unknown provenance with the realization that the answers may never be found. It is almost like DNA came from the prior Universe and has interfaced with RNA here and together life has formed.

It could be that DNA predates the present Universe that we are in. I have thought so when realizing how novel the molecule is. It could be that perhaps a signal was transmitted across the beginnings of our Universe and it could have influenced the formation of molecules, etc. It is through DNA that we seem to violate the laws of Entropy, and so thusly DNA may be the result of an intelligent signal transcending our Universe.

Mystery extra-galactic radio bursts could solve cosmic puzzle

Ultrashort radio bursts from outside the Milky Way may help locate missing baryons.

NATURE

Ron Cowen

04 July 2013

"Astronomers have for the first time detected a population of ultrashort radio bursts with properties that strongly suggest that they originate from outside the Milky

Way Galaxy. Lasting for a few thousandths of a second and estimated to erupt roughly every 10 seconds, the mysterious bursts are likely to be caused by a previously unknown class of radio-emitting phenomenon, researchers report in Science"

"…"This is one of the most important radio discoveries in the last couple of decades," says Scott Ransom, an astronomer at the US National Radio Astronomy Observatory in Charlottesville, Virginia, who was not part of the study."

Thusly: The laws of entropy are not violated with DNA because it has arrived from another Universe and so the laws of thermodynamics are conserved after all. DNA and RNA based life has found a loop-hole in the laws governing the physics of everything. At first what seems counter intuitive suddenly make sense and the realization of knowing the profound mystery has been realized as the laws of thermodynamics are conserved.

Furthermore: the Universe prior to ours could have had slightly or vastly different constants so the formations of the DNA templates would have formed under different circumstances. The surviving artifacts would then be in a sense cheating death of the previous Universe and could have arrived here impervious to some of the downfalls of our Universe.

There could be a situation where the dating of the Universe is wrong also. A recent discovery of a star that has anomalous readings could indicate that what we believe happened at the dawn of our Universe is completely wrong.

Oldest Star Known Is 14.5 Billion Years Old, Very Close to Earth

Softpedia

March 8th, 2013, 14:11 GMT

By Lucian Parfeni

"NASA is revealing its findings on the oldest star we know of, so old that it precedes the Universe…"

"The Big Bang happened 13.77 billion years ago. But there's a margin of error on the age of the star of plus or minus 0.8 billion years, so at the edge it marginally fits inside the timespan of the universe."

"That's actually a big improvement over earlier measurements which said that the star was 16 billion years old."

"In fact, that's exactly what prompted astronomers to look even closer at the star, HD 140283, and try to determine its age more accurately."

> *"The discrepancy in the estimated age of the star and the estimated age of the universe means that there's an error somewhere, either in the way the age of universe is calculated (unlikely) or in the way the age of stars is calculated. Alternatively, it could mean the star is closer or further away than we thought."*

In the thoughts of the Cosmic Ray Background Radiation, and our readings of it, and the age of the Universe there could be awaiting discoveries to be made. Since there was a supernova in the past in the local neighborhood here: there could be a case made that we are living in the shockwave remnants of a previous supernova and the cosmic ray background radiation is simply that shockwave.

What we actually think of as the beginning of the Universe could simply be the starting of the supernova that caused our sun to condense and form an accretion disk. It would seem very arrogant of us to believe that we have found the real ends of the Universe itself with the cosmic ray background radiation.

In fact the Universe appears to us in the remnants of 23 percent dark matter and 73 percent dark energy while DNA is the yardstick of the illumination of the hidden. DNA tells us that there are hidden aspects to the rest of the Universe perhaps parallel to us as indicated by Quantum Mechanics where besides the space in the atoms there are other realms where the electrons dart in and out of revealing to us the rest of the real

Universe. The novelty of DNA tells us that the Universe is hidden in plain sight just beyond our detection.

New Light on Dark Matter

Science 20 June 2003:

Vol. 300 no. 5627 pp. 1909-1913

Jeremiah P. Ostriker, Paul Steinhardt

"Dark matter, proposed decades ago as a speculative component of the Universe, is now known to be the vital ingredient in the cosmos: six times more abundant than ordinary matter, one-quarter of the total energy density, and the component that has controlled the growth of structure in the universe. Its nature remains a mystery, but assuming that it is composed of weakly interacting subatomic particles, is consistent with large-scale cosmic structure. However, recent analyses of structure on galactic and sub galactic scales have suggested discrepancies and stimulated numerous alternative proposals."

There is a relationship to the human genome and the diameter of the solar system, the distance to the sun, and even the distance to the moon. There is a resonance to the length of human DNA and the solar influence on the surrounding galaxy. Thusly: The length of DNA of a human being seems to be related to the wavelength that corresponds to the diameter of the solar system and/ or other astronomical constants.

The age of the Universe plays a role in thoughts of where DNA has come from in the first place. The structures of it are so novel that it has Transcendence properties to it. DNA is so fantastical that it rivals the Universe itself when contemplating it. DNA gives us an idea of how the real Universe underneath light and dark matter must operate since the molecule is so vast.

The Pre-biotic Viral World

Virologists also have a clique theory among them that the viruses are a remnant of an ancient sphere of a prebiotic, protoplanetaria world. Perhaps planetesimals were or are covered with viruses that are nonliving structures. These views are the result of the way the viruses react, and respond and function. The viruses would be pioneer debris forming and their structures form in a pre-biotic world. The viruses that are not alive at all have prototype organelles inside that manufacture protein as well as conduct activity with the help of enzymes.

Viruses have the ability to survive in harsh environments as propagules even to the extent of x-ray bombardment as the first research demonstrated when the photographs were taken of the unknown materials thought to be viruses. In such thought of an RNA world of viruses the RNA uses polymerases to synthesize proteins yet the viruses are not alive. Mechanical artifacts would thusly arise as viruses which have the ability to replicate.

The nature of the viruses is that they can have a shape of a crystalline structure and be virtually indestructible. They give baffling results to efforts at decoding, categorization, and explanation. They only have one purpose with no superfluous machinery and that is to consume their hosts and reproduce. Viruses do all of their activity and yet are not alive.

Thusly viruses are mechanical fragments in the cosmos wreaking havoc upon everything in their paths as they survive in extreme environments waiting to be uncorked like champagne bottles. The pressures of virus genetic materials are measured in terms of pressures of that of about a champagne cork literally. They are like jack-in-the-boxes waiting to activate.

In terms of what drives a virus, and or what makes it go so to speak it is a combination of mechanical energy stored in the package and a shape of the various components that cause chemical reactions to take place. These behave as metaparticles with information that is in a state of dynamic flux able to be programmed where in actuality most of them in reprogramming do not survive at all. The heat above absolute zero, the stored spring like mechanical energy, and the shapes of the devices are what makes the viruses do what they do.

A recent article in Science News indicates that a recently discovered large virus may hold clues to the natural science of the first viruses:

News in Brief: Size isn't only mystery of huge virus

Strange replication method and unusual genetic sequence among the mysteries

Science news

By Cristy Gelling

Web edition: July 19, 2013

"The largest virus ever identified has been found on the seafloor off the coast of Chile. Pandoravirus salinus is about twice as long as the previous record holder, Megavirus chilensis, with a genome that is twice as large. That makes P. salinus larger than the smallest bacteria."

"The authors suggest a controversial hypothesis for why the Pandoravirus is so odd: It could have evolved from a type of free-living, ancient cell that no longer exists. Its discovery is likely to add fuel to the heated debate about the evolutionary origins of viruses."

Thusly: a precursor to life is housed in the structure of at least one virus, with all of the clues as to what went on in the distant past.

The Antigenic Shifting in Influenza

The comparative case for Panspermia could be thus: a 5 epitope point mutation in antigenic shift for an influenza virus has a probability of $P = (1/10^6)(1/10^6)(1/10^6)(1/10^6)(1/10^6) = (1/10^{30})$. This indicates that for a change to occur in all 5 epitopes at the same time for an influenza virus the P (probability) exceeds the number of particles of influenza virus on Earth yet there are pandemics of influenza happening all of the time. There is the 5 point epitope shift when two viruses combine the other's epitope shifts. This is comparative to the way life must have originated on primordial planet Earth. The comparative case for Earth Panspermia by the numbers suggest an input from another source: the same as the case for the 5 point epitope Influenza virus shift as it gets an input from another source.

Figure 3. In this drawing: an alien possibly an unknown victim of viruses in the cosmos distributing them to us in the Galaxy.

Perhaps viruses are surviving artifacts. There must be unknown hosts to harbor and provide substrate for the viruses to propagate. Space may be covered in abiotic viruses floating around from prior civilizations and activities of the processes of life.

There should be a theory where an unknown species would be the host for a progenitor of basal viruses where at that location in the Universe through means unknown the viruses are spewn out perhaps by traumatic volcano eruption, planetary destruction or by way of supernova which is the result of each star's life cycle. The basal viruses would then arrive like coconut seeds on a volcanic island. They would be the pioneering viruses in a habitat of fresh hosts spreading the genome into the biosphere where it would mutate and abiotically adapt and generate more propagules.

Denisovan, Neanderthal Viruses Discovered in Human DNA

Nov 20, 2013 by Sci-News.com

Scientists from the University of Oxford and Plymouth University, both in UK, have found evidence of Neanderthal and Denisovan viruses in DNA of modern humans.

"In 2012, researchers from Albert Einstein College of Medicine identified remnants of 14 ancient viruses in the genome sequences of Neanderthal and Denisovan fossils, dating back about 40,000 years ago. But they failed to find remnants of these viruses, belonging to the HML2 retrovirus family, in the human reference genome sequence."

"In a new study, Oxford University researcher Dr Gkikas Magiorkinis with colleagues compared Neanderthal and

> Denisovan data to genetic data from modern-day cancer patients and managed to identify remnants of one Neanderthal and seven Denisovan viruses."
>
> "The discovery will enable scientists to investigate possible links between HML2 retroviruses and modern diseases including HIV and cancer."

Look at this quote! It suggests that viruses from another species infected us or perhaps created us in the first places. We may have been mutated from a Neanderthal by this very virus. It could be that viruses are the hidden hand creating new species.

The operations of some viruses are that they insert parts of their sequences into their hosts using the host machinery in their cells to replicate themselves. Viruses carry these foreign genetic materials in their structures as the sequences. Thusly: viruses assemble portions of artifact genes in each new host. It may be the case that virus propagules are building foreign aliens on planet Earth when they replicate.

These foreign viruses are slowly wrenching us mechanically into ancestral extraterrestrials that may have lived millions of years ago in another part of the Universe. The viruses drive evolution and may be the hidden forms at the basal trees of all life. Notice how there is never really basal forms at the branch of each speciation event but a hint at a mechanism only. Perhaps everything looks the same after so many millions of years spanning the entire Universe regardless of origin.

Viruses force the hand of a card playing deck consisting of the amino acid sequences of RNA and DNA of the host where some of the types of viruses insert their code in places still being researched today. We are all helpless and get dealt these hands and play the virus game trying to survive the hands of time as it is inevitable – we are at the complete mercy of viruses and their destructive wake.

The Comet Debris

A recent probe sent to a comet has discovered the nucleotide Glycine flowing out of the debris of the comet. The probe was flown right through the tail of a comet, and gathered samples in the aerogel a new high tech substance. When analyzed the tell-tale Glycine was found stuck in the aerogel.

Organics Captured from Comet 81P/Wild 2 by the Stardust Spacecraft

Science

**15 December 2006:
Vol. 314 no. 5806 pp. 1720-1724
DOI: 10.1126/science.1135841**

"Organics found in comet 81P/Wild 2 samples show a heterogeneous and unequilibrated distribution in abundance

and composition. Some organics are similar, but not identical, to those in interplanetary dust particles and carbonaceous meteorites. A class of aromatic-poor organic material is also present. The organics are rich in oxygen and nitrogen compared with meteoritic organics. Aromatic compounds are present, but the samples tend to be relatively poorer in aromatics than are meteorites and interplanetary dust particles. The presence of deuterium and nitrogen-15 excesses suggest that some organics have an interstellar/protostellar heritage. Although the variable extent of modification of these materials by impact capture is not yet fully constrained, a diverse suite of organic compounds is present and identifiable within the returned samples."*

The Outer Solar System Processing Plant

What could be happening is that in our case of solar system dynamics organics are formed from another place such as that on the moon of Saturn Titan where entire oceans of organics are currently present where the statistics are better for organic formation and from there the organics are spread to other parts of the solar system by bombardment processes of the asteroids. An organic genesis could be on the moon Titan where then they are ready for the next step in a processing plant of our solar system as a whole system.

The processing plant for the organic synthesis could be the dynamics of Saturn, its moon Titan, and the magnetic processes as noted by the ring structure of Saturn. The outer planets could be the missing part of a processing plant where the statistics seem to indicate another source for manufacturing of the basic building blocks of life to occur: there are oceans of organics on the moon Titan ready for transfer to the next step.

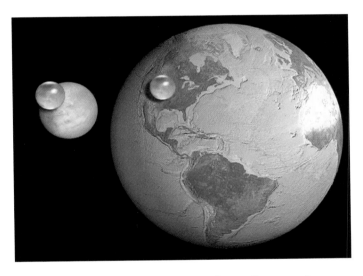

Figure 4. Earth - Europa Water Comparison (NASA/ JPL)

In an organic example: cyclohexane looks like the molecule glucose with a few additions. There is a runaway life process for plant life with a ring structure using magnesium as the basis for chlorophyll and photosynthesis by plants polymerizing glucose to make cellulose. It is all based upon the manufacturing of organic molecules of glucose ready to be formed in to cellulose.

The solar system in our case, and all over the Universe, must function together as a whole unit and not an isolate of the habitable zone of any one place. This would thusly also constrain the search to further refinements and thusly chip away at the number of uninhabitable planets in the Universe revealing habitable systems. There must be a vast array of planets working in concert to form life and organic molecules in each solar system and also local neighborhoods of stars.

These components of life obviously have come together and have formed basic life that falls to the Earth measured in the tons per day. A recent article has been published describing these organisms. It might be determined that these forms are a type of proto-life.

The truth IS out there (above Cheshire, that is): British scientists claim to have found proof of alien life

Life on Mars? No, say scientists, it's floating 27km above Chester

The Independent

Thursday September 19 2013

"A team of British scientists is convinced it has found proof of alien life, after it harvested strange particles from the edge of space."

"The scientists sent a balloon 27km into the stratosphere, which came back carrying small biological organisms which they believe can only have originated from space."

"Professor Milton Wainwright told The Independent that he was "95 per cent convinced" that the organisms did not originate from earth."

"..."By all known information that science has, we know that they must be coming in from space," he said. "There is no known mechanism by which these life forms can achieve that height. As far as we can tell from known physics, they must be incoming."..."

The Meteorites

The first case in historical documents that I found that involved finding something in meteorites was from the 1800s. They weren't sure what had been found. It made the rounds in haute society at the time in their literature. I can infer that they were observing fungus.

Figure 5. In this photo the author shows his research in to coral fungi – unusual organisms that present tantalizing characteristics: one being that they don't use chlorophylls. These things will be the organisms found all over foreign planets. (Author).

Meteoric Organisms

Knowledge

Jan 1882.

Carl Vogt.

> *"The organisms in meteorites announced by M. Hahn, have no existence…None of these imagined organisms has the microscopic structure belonging to the organisms with which they have been associated. In particular, the asserted sponges do not show the structure either of existing or fossil sponges…"*

Look at this: it is the smoking gun. This is the Holy Grail of fossils in meteorites. I read several accounts of the descriptions from the 1800s. They seem to be describing coral fungi but didn't know what they were. One reference claims that the organisms were coral fossils. Then they go on about sponges. Later the evidence began to stack up with meteorites.

Meteorites and Planetary Organic Matter

Observatory

82:216-218, 1962.

Briggs, Michael H.

"….the recent discovery of petroleum-like hydrocarbons within carbonaceous meteorites has raised an interesting problem. Either the meteorite hydrocarbons are part of the remains of an extra-terrestrial life-form, or they are abiotic compounds formed in space. …raises the question of whether any of the organic constituents of terrestrial petroleum are compounds brought to earth by meteorites."

Biological Materials In Meteorites: A Review

Science

151:157-166, 1966.

Urey, Harold C.

"...as briefly reviewed above, strongly suggest that biogenic materials exists in these meteorites and that it may be indigenous."

Fossils have been found in meteorites since antiquity. Recently one was analyzed in Britain. The accounts are evenly dispersed through time. Some one finds the materials, publishes and then nothing happens.

Meteorite Holds Proof of Extraterrestrial Life — Or Not

Live Science

Date: 23 January 2013 Time: 03:17 PM ET

Marc Lallanilla, Life's Little Mysteries Assistant Editor

"A British professor of astrobiology has asserted in breathless tones that a meteorite found in Sri Lanka contains microscopic biological fossils —

indisputable proof, he claims, that life exists beyond Earth. Other scientists, however, have cast doubt on his claim."

"Professor Chandra Wickramasinghe, director of the Buckingham Centre for Astrobiology at the University of Buckingham in England, states in an article from the Journal of Cosmology that diatoms — a type of microscopic algae — found in the meteorite are extraterrestrial in origin, the Huffington Post reports."

In another source there appears to be a reference to the same incident:

We are not alone: 'Alien life' discovered in meteorite which crash-landed on Earth

The Mirror

23 Jan 2013 00:51

"A top British scientist discovered the two-inch wide rock was pitted with tiny fossils of algae, similar to the kind found in seaweed"

"A top British scientist claims he has found proof that extraterrestrials exist after cracking open a meteorite."

"Instead of finding an alien like Hollywood favorite ET, Professor Chandra Wickramasinghe discovered the two-inch wide rock was pitted with tiny fossils of algae, similar to the kind found in seaweed."

"The respected professor believes it proves we are not alone in the universe."

"He said: "These finds are crushing evidence that human life started outside Earth."

"The rock was one of several fragments of a meteorite which crash landed in central Sri Lanka in December."

"They fell to earth in a spectacular fireball and were still smoking when villagers living near the city of Polonnaruwa picked them up."

"The fossils were discovered when the rocks were examined under a powerful scanning electron microscope in a British laboratory."

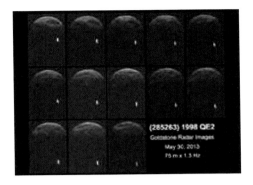

Figure 6. Asteroid 1998 QE2, May 2013 Flyby (NASA)

Something occurred with a recent meteorite observed by NASA radar. In May of 2013, a near miss asteroid passed by the Earth. The interesting thing about it is that it had an orbiting body around it. The asteroid does not calculate to have enough of a G to allow for such a suborbital object around it. This is a smoking gun of something so unusual that it is extraordinary. The velocity of the asteroid with the suborbital object around it was that of a gunshot, and so it is forever lost in the solar system beyond our grasp.

Life on Mars

Figure 7. Planet Mars. (NASA/JPL)

While speaking at a meeting of the American Geophysical Union conference in San Francisco, lead scientist John Grotzinger of the California Institute of Technology indicated the mission of NASA's Mars rover Curiosity was to search for traces of life on the Planet Mars and – it has been successful in finding traces of compounds that contain carbon. There is also evidence to believe that there used to be water at the landing site of Curiosity where it touched down in August-2012. Scientists had expected to find more evidence of organic compounds as the Mars rover continued to explore the sands towards Mount Sharp on its mission to start digging deeper. On Earth NASA also announced at the fall 2012 meeting of the

American Geophysical Union (AGU) that the solar powered rover Opportunity that had been in operation for 9 years had found a source of usable neutral water in the form of bound clay materials in the Endeavour crater on Mars.

Mars Ecosystem - The Same Biome, Everywhere.

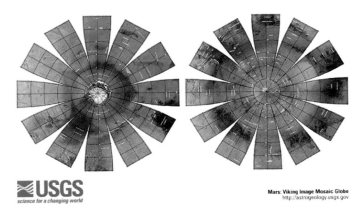

Figure 8. In this photo: The Viking Mission Map (NASA/ USGS). Prospecting around, ready to go – Mars.

The "Canals" of Mars were among the first mysteries observed from our Earth. They turned out to be the optical illusions of the major features of Mars when viewed close-up, such as the apparent great X on one plane of Mars. The Viking Mission resulted in robotic equipment exploring the surface with analysis of the substrate that changed the perception of the planet itself. The results of the current Spirit and Opportunity scientific exploration resulted in some photographic landscapes that are alien to our world in the respects of the macroscopic features and microscopic features. It is ironic that there is under serious discussion the subject of Mars having water features of past erosion from long forgotten epochs.

 On day 699 of The Spirit Rover Mission, and Day 679 of The Opportunity Mission, to the planet Mars the one striking if not shocking aspect to the photo analysis of these pictures is that Mars is the same biome all over the planet. This could have been any day of this mission for this examination. The dust storm that was global just prior to these landings had faded, and aside from a few dust devils the "weather" had not changed during these mission pictures, but The Viking mission had actual weather changes from day 28 of the Viking mission to day 615. These recent rover missions have enjoyed good global over-all average weather: in essentially the same biome.

Figure 9. Opportunity Day 679

Figure 10. Spirit Day 698

Perhaps our variation, of Earth climate measurements, are in a chaotic cycle of equilibration of temperatures, and pressures as compared to the apparent static nature of Mars: Mars is a study in Static conditions as compared to Earth. The small stones give evidence and testimony to the current power or lack of it of wind: The drifting sand dunes, and pebbles in the Opportunity picture can give an idea as to the actual wind force. The Sprit picture actually looks like more of the remnants of past water power as a moving force than of the wind force. Yet these long periods of Statics were not always so, as with the case of some of the observations from satellite reconnaissance that witnessed global dust storms on Mars, which may indicate that the weather may not be controlled on all planets by the local Sun, but by more galactic forces itself, or other forces that we can only speculate on - Mars has an indication that its tilt as the same as our tilt may be controlled by our solar orbit in the Galaxy itself: The current static nature on Mars actually indicates that more weather aspects of our two planets are controlled by our orbit around the Galaxy, rather than our distance from the sun; however the current conditions on Mars are static for the missions of Opportunity and Spirit as well as Curiosity.

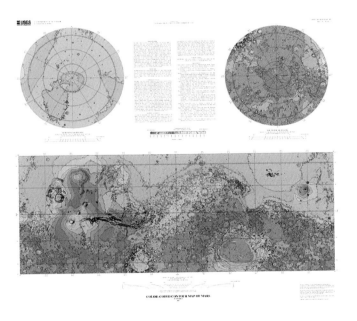

Figure 11. Topographic Image of Mars

The Dynamics of Mars is a function of something in the past. There may have been water in the Dynamics of Mars in unknown epochs. From The Viking Missions to the Spirit and

Opportunity Missions, there suggests that something apparently happened in the remote past there involving large bodies of water. These images from orbiting survey platforms show a larger scale of planetary formation, possibly, and likely involving large standing bodies of water over vast amounts of time.

The current conditions of Mars are that of a static desert, where the weather changes are slower than on Earth and may actually be slightly dynamic due to the orbit and location around the Galaxy, something that we can learn from visiting such a place. Water seems to have played a part in shaping some of the macroscopic features at some point in the past there.

The Findings of Mars

There was a recent conference at UCLA on the habitability of Mars. The UCLA Institute for Planets and Exoplanets, The UK Center for Astrobiology and the NASA Astrobiology Institute conducted a two-day conference in February 2013 that examined the present-day habitability of Mars. These institutions recorded the proceedings where various media files can be streamed from their websites.

Mars may have a fossilized organism that can be identified by pictures. The slime mold Myxomycete can be identified in at least two life stages in either the endosporangial form or the exosporangia form along with the amoeba stage.

In one photo the ameba life stage is seen, and in other photos the sporangia stage has been apparently photographed as well.

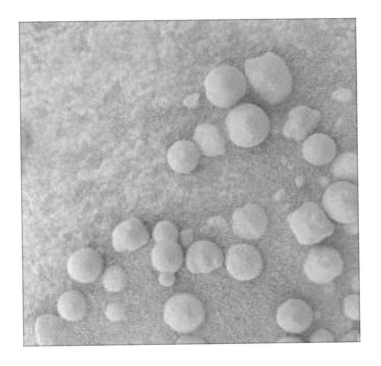

Figure 12. Apparent Fossilized Myxomycetes Sporangia Stage. Credit: NASA/JPL

The exosporic myxomycete, and/ or endospore myxomycete photographs can explain a lot of puzzling fossils discussed in popular literature where the authors couldn't explain the type of fossilized organisms found in meteorites from the 1800s (see end notes).

The "blueberries" photographed on Mars on March, 3rd 2004 by the Opportunity rover can be explained only in context when taken together with the other picture of the apparent amoeba stage Myxomycete.

Figure 13. Apparent Myxomycota Ameoba Stage Fossil. Image Credit: NASA/JPL-Caltech/Cornell University (NASA/JPL).

The slime mold in the amoeba stage photograph is the smoking gun of all Mars missions:

NASA Mars Rover Opportunity

"This image from the panoramic camera on NASA's Mars Exploration Rover Opportunity shows a rock called "Chocolate Hills," which the rover found and examined at the edge of a young crater called "Concepción."

"The rover used the tools on its robotic arm to examine the texture and composition of target areas on the rock with and without the dark coating. The rock is about the size of a loaf of bread. Initial analysis was inconclusive about whether the coating on the rock is material that melted during the impact event that dug the crater."

"This view is presented in false color, which makes some differences between materials easier to see. It combines three separate images taken through filters admitting wavelengths of 750 nanometers, 530 nanometers and 430 nanometers. Opportunity took the image during the 2,147nd Martian day, or sol, of the rover's mission on Mars (Feb. 6, 2010)."

The problem with finding such an organism traditionally studied by mycologists although do not represent true fungi is that organisms drive everything out of their way in to extinction, even if it means completely destroying their planet. What we may find in the Universe when we go exploring is planet after planet driven to extinction events. The myxomycetes may have killed everything on the planet Mars.

This can be proven as easily as visiting your neighborhood pet store and discussing salt water fish tanks with any of the sales staff. The perils of a salt water tank are watching a wild kingdom ensue where one of the species destroys themselves and others while attempting to secure dominance of the environment. The salt water tank trick is to try and get everything in balance so that of course there could be attempts at watching a peaceful tank.

What's even more intriguing about the finding of apparent fossilized slime mold(s) on Mars is the fact that they appear to be on a fossilized timber – it even looks like a square timber from an ancient pirate ship – no kidding.

The most obvious source for the fossilized square wooden timber is this: the volcanized explosion from Earth in 1450 b.c of the island of Thera that destroyed the entire Minoan military naval fleet. One of the center boards of a Minoan naval warship was blown all the way to planet Mars. This would explain where the slime mold came from. It's Panspermia of Mars from planet Earth.

Identification although based upon fossil remains may be more unlikely because the organism uses chitin is that the organisms could be Ascomycota. There could also be definitive candidates for Basidiomycota as well – thus the core requirements of being a microbiologist are therefore mycology. The fossilization process on Mars would have to be derived to determine exactly what it is we are looking at.

There would be a new definition of fossil on Mars due to the process being different as compared to Earth fossilization. Once the definition of fossil on Mars could be determined then the next step would be to determine exactly what an organism from Earth would do in the absence of the Earth atmospheric conditions present on Mars. Then the next step would be to determine what an organism would look like as developed on Mars but due to the conditions that occurred on Mars – what would it look like if it evolved on Mars and then died on Mars after an apparent global tragedy there. Each case would determine a new definition of fossil.

The isotopic ratios of the fossils would need to be investigated to determine from what planet of origin the apparent organisms came from. After the isotopic ratios results came back from analysis then it would need to be determined at exactly how long the organisms spent on each planetary body.

Figure 14. Current possible example of Mars life. Image Credit: NASA/JPL-Caltech/Cornell Univ. /Arizona State Univ.

NASA

2014-01-21

"This before-and-after pair of images of the same patch of ground in front of NASA's Mars Exploration Rover Opportunity 13 days apart documents the arrival of a bright rock onto the scene. The rover had completed a short drive just before taking the second image, and one of its wheels likely knocked the rock -- dubbed "Pinnacle Island" -- to this position. The rock is about the size of a doughnut."

"The images are from Opportunity's panoramic camera (Pancam). The one on the left is from 3,528th Martian day, or sol, of the rover's work on Mars (Dec. 26, 2013). The one on the right, with the newly arrived rock, is from Sol 3540 (Jan. 8, 2014). Much of the rock is bright-toned, *nearly white. A portion is deep red in color. Pinnacle Island may have been flipped upside down when a wheel dislodged it, providing an unusual circumstance for examining the underside of a Martian rock."*

"The site is on "Murray Ridge," a section of the rim of Endeavour Crater where Opportunity is working on north-facing slopes during the rover's sixth Martian winter."

There is a picture comparison that shows something like a biotic form on Mars. It looks like a lichen form but with an unclear shot it is impossible to tell. The capacities of lichens which comprise two organisms are that they dissolve solid rock. The picture could indicate that Mars is a powder keg waiting to explode with life.

For the lichen there must be a symbiont, in which co-exists with the fungus which could be a cyanobacterium or a green algae. The fungus part of the lichen I am on the trail of for Mars with several notations from sources. We may be looking at fungus and a photobiont of something we are not familiar with.

Pigments for photosynthesis and for Mars organisms may be different so there might be a case made for purple pigmented alga something that may be rare or even nonexistent on Earth in the slightest. The Lichen in the picture may be a purple pigmented alga and the Mars fungus or even a triple organism of the previous two indicated with a form of stable bacteria.

Scratching around up there has undoubtedly released spores of bacteria that have whipped around Mars several times for each scratching event and they have mutated in the atmosphere of Mars mixing with the established fungus on Mars.

Figure 15. Image: Ancient Mars with ocean (NASA / GSFC)

"'We are all Martians': Chemist's otherworldly claim stirs debate"

Alan Boyle, Science Editor NBC News

Aug. 28, 2013 at 7:32 PM ET

"This graphic shows how Mars might have looked billions of years ago, when scientists believe the planet had a large northern ocean."

"Are we all Martians? A controversial hypothesis contends that life on our planet

had to get its start somewhere else — most likely on Mars — because the chemistry on early Earth couldn't have provided the required molecular machinery."

The obvious dilemma is that the Mars missions are considered automatic primary literature, and the data must be used directly for writing scientific research. Any hypothesis would be oriented to primary literature. Since the results would have to be repeated it leaves a conundrum of having to repeat the tests by another research organization – a sheer impossibility at this time. The write ups for these hypotheses would also be multidisciplinary something not usually taken on by anyone. So the write-ups are unusual and the testing cannot be duplicated.

The write up would be: Although the structures that look like organisms appear to be natural mesomorphs they may be fossils, etc. There would be the opposing argument and then the presentation that there was life and then fossils, and so forth. The write ups would be tantalizing and then fall in to the rest that can't be tested and so on.

In mathematics there is presented a problem, and then in Cartesian systems there are the solutions and some have two roots. What if the problem really doesn't give an answer that matters? We must state the problem in other terms.

Mars looks like at one point it had a tremendously dynamic atmospheric disturbance on the surface and at one point had one or more organisms that are comparable to Earth biota. Mars is the obvious planet of choice for any Panspermia investigation because of the sheer fun factor involved, and because the real estate is so much more conducive to exploration as compared to the asteroids or the Moon, and because it looks like as the investigations continue a much clearer picture has been emerging with a lot more agreement on what has happened.

Life on Earth

Figure 16. Yellow Stone National Park: Current Prokaryote Life in Primordial Conditions.

Since animal life is so complex it begs the question: where did the gene(s) that started animal life and eventually human life and speech come from?

In basic terms there are clouds of amino acids in space. These nucleotides have been polarized by a close pulsar and have the same twisting nature. Notice how all life as we know it here has DNA that twists in the same direction. They are self-assembling molecules in the same orientation.

The characteristics of light are that going in one direction is the "e" field portion of the light wave then perpendicular to the "e" field is the "b" field of the propagation, thereby the pulsar would polarize the nucleotides found in our unknown nursery of the cosmos that we haven't found yet but are certain must exist since the molecules are in the same direction.

Structures present in the Milky Way suggest that the spiral arms may act in a certain polarized way that was discussed in the primary literature. This would polarize molecules.

Rotation Measures of Extragalactic Sources Behind the Southern Galactic Plane: New Insights into the Large-Scale Magnetic Field of the Inner Milky Way

Astrophys.J.

663:258-266,2007

"... We show that the magnitudes of these RMs oscillate with longitude in a manner that correlates with the locations of the Galactic spiral arms..."

Thusly one of the mysteries of the amino acid building blocks can be put to rest: the polarization of our neighborhood amino acids are manifested by at least one and possibly more ways.

Our sister stars that were formed at the same point in the rotation of the galaxy should be nearby, and blowing out amino acids, and base materials while we rotate through them. We may even be rotating through slightly more advanced or extremely more advanced clouds of stuff.

Billions of years ago life existed in the form of single celled organisms on planet Earth. The analyses of the G to C ratio suggest that primitive life forms were fine and would have been happy until all the oceans evaporated like on Mars if not for an event that disturbed the life forms at the time when an unknown sequence arose. The event was the famous gene that created the first animal. Then along arose fish that eventually moved onto land and then appeared the hominids and out of nowhere the gene that encoded language in human beings. Was the arrival of life, speech and intelligence on Earth due to the events of Panspermia?

There are complicated multi-celled lives now that have speech centers in their brains. Using microbiology we can trace the origin of life, animal life, and the rising of intelligent life based upon the universal speech processor.

Figure 17. In this image: Prokaryotes conjugating foreign genes. E coli have taken up the jellyfish gene for bioluminescence. (Author).

A gene came out of nowhere and created the first animal approximately 600 MYA according to fossils. How could it be? There is the fact that bacteria incorporate genetic material outside of cells into their genome that could explain it. If you crush up genes and place them in a petri dish the organisms take up some of the genes. Bacteria like E. coli take up genes in their environment and by transformation, and conjugation they attempt to incorporate them into their genomes. If you have a petri dish full of E. coli and you place a bacteria with a gene that glows in the dark it soon spreads to the others and they all glow in the dark due to the conjugation (See appendix). The gene that enabled animal life spread to other organisms in the same way using the property of conjugation.

A series of adaptations occurred from the prokaryotes in primordial Earth. The adaptations resulted in groups so different they are referred to as separate kingdoms. Some primary literature sources refer to a total of 17 separate kingdoms at the level of basal organisms in between prokaryotes and the plants.

- Amitochondriate groups
- Heterolobosa
- Physarum
- Euglenoids
- Amoeboflagellates
- Dictyostelium
- Red Algae
- Stramenopila
- Plants
- Fungi
- Animals

The smoking gun of Panspermia is in the details of the machinations of all of the life in between single celled life and multi celled eukaryotic life. In those life forms are the clues to Panspermia. Entire planets should basically be just one kingdom since all life would be thought to start in one place or perhaps should be a couple of basal kingdoms.

Planet Earth Biota

Intelligent Animals on Earth arose from single-celled eukaryotic ancestors. Characteristics of animals include:

- Multicellular body plans

Heterotrophic nutrition by ingestion
Ability to move, at some stage, from place to place under their own power
Embryonic development from zygote to adult
Presence of flagellated cells in some stage of their life cycle

The First Animal Blood Cells

Deep in epoch antiquity of animal precursors is the story of how the first blood cells originated. Blood plasma to this day is almost exactly like sea water constituents. Circulating blood cells of today are bathed in the artificial sea water by developed hearts of the many celled and tissue typed animals.

To really find the concentrations of primordial sea water today one has to look in the inner ear of an animal to get the exact constituents. This brings to mind all kinds of ideas of testing of extraterrestrial life, primordial life, and places the story of how life evolved in to perspective. In other words although blood plasma appears to be somewhat like sea water the smoking gun of this line of thought is only really found by examining the inner ear: animal plasma is not as close to primordial sea water as the fluid of the inner ear.

Observing the maturation of modern blood cells there are clues to the evolutionary story of how they must have originated. The stem cell precursors give rise to certain lineages of types of cells. Even stem cells at a certain point are indistinguishable between red cells and white cells.

At some point the mitochondria became incorporated into all animals to enable efficiency of energy utilization. The primordial soup thusly contained mitochondria, white blood cells and red blood cell precursors. Animals engulfed mitochondria into their makeup to incorporate mitochondria as symbiotes.

What is fascinating is the relationship with mitochondria an organelle that was once an independent organism and the red blood cell. Mitochondria finalize the step of hemoglobin manufacturing for incorporation in to red blood cells, throwing a wrench in to the works of exactly knowing what went on in the primordial soup.

There is a huge conundrum of how and why the first red blood cell developed in relationship to the mitochondria. Yet there could not be a situation where each can be separated from the other. What is known is that there was an interaction.

The animal precursor examples: red blood cells, hemoglobin, and mitochondria that can be observed in transitory positions are long ago lost in antiquity. There must have been animals with hemolymph, mitochondria and red blood cell precursors. Today there has not been found an animal with the characteristics of both hemolymph and red blood cell precursors.

Interacting with the membrane and the receptors there is also a lattice structure of proteins in the red blood cell that hold the cell together. In the red blood cell the proteins are made out of spectrin and form a matrix for the cellular components to reside in. The mature red blood cell in the basic form is more than a membrane but also contains proteins that provide a skeletal structure thus adding to the complexity of even the most basic of cells descriptions.

To really nail down the first animal cell and the first blood cell there has to be a diagramming and flow charting of the interactions of what lead to what. It could take some doing to trace down the entire complete picture.

The various blood cells, and mitochondria were gathered together in the first animal and became symbionts to form structures in the oceans that were more than 2 cells thick. Axioms of inheritance enabled progeny to change morphology. The way that animals accomplished the task was by storing the code for a program in structures called DNA.

Looking at white blood cells in the lineage of macrophages one can easily see how they could have existed on their own in a primordial soup eating bacteria by themselves because that is their job in the animal physiology – engulfing foreign cells.

The Arthropods

The insect life forms accomplish hemoglobin circulation with the use of red blood cells. The hemolymph originated from the first life in the oceans. It is with hemolymph that animals began to develop into large assemblages. With hemolymph structures could be many cells thick. Hemolymph may hemoglobin but is not found with a large amount of red blood cells, or may be found with an absence of red blood cells and hemoglobin.

The hemocytes in insects are platelets, and white blood cells, and in some aquatic insects such as the midges there are hemoglobins. Some hemolymphs developed to provide a type of antifreeze to the animal. This gives the first clues as to how and when blood and the precursors began.

There would be an entomologist, a hematologist, and an evolutionary biologist in a hypothetical study of these affairs. At some point all of the orders of insect precursors had not only hemoglobin, but red blood cells, but lost them in revisions as they moved on to land.

The insects don't require closed circulatory systems with red blood cells containing large amounts of hemoglobin because they can respire through their exoskeletons directly and as invertebrates they don't have a large nervous system.

A research on insects can reveal references to no hemoglobin present, to some hemoglobin present, to no red blood cells present to some red blood cells present. What is accepted is that insects have hemolymph and may be the only animal precursors worth studying in these lines of thoughts. It may be that the entomologists have examples lying around with code snippets in the aquatic forms.

Once complex life established itself on this planet, in whatever mannerism, the next steps became exploiting the resources on land. Arthropods came out of the oceans to exploit land. They were perhaps first on the scene.

> *"The great radiation of modern insects began 245 million years ago and was not accelerated by the expansion of angiosperms during the Cretaceous period. The basic trophic machinery of insects was in place nearly 100 million years before angiosperms appeared in the fossil record.(Labandeira, C., & Sepkoski, J.,1993)."*

Insects then began radiating once established on land. The insects dominated for millions of years. They were making a living on land before the angiosperms took hold, so they must have exploited the gymnosperms prior to the angiosperms.

Mars shows signs of nuclear blasts 180 million years ago, so if it was the result of an organism from Earth it would have to be an intelligent Arthropod life form that no one has ever seen before.

Ever wonder why the red planet is red?

"About 180 million years ago, a planet-shattering yet naturally occurring nuclear reaction may have wiped out everything on Mars, sending a shockwave that turned the planet into dry sand (Brandon, 2012)"

Since Anthropologists anthropomorphize everything into hominids – I imagine an unknown arthropod hominid out there in the cosmos waiting to meet us.

Figure 18. In this image: Artist drawing of an unknown arthropod life form using technology.

If there were intelligence development before our present age of hominids it would have to have been insectoid life- the Arthropods. There has not been any large Arthropod life discovered that used anything other than breathing through pours of their exoskeleton and also using hemolymph. Thusly: the limitations of Arthropod life are huge confining them to small forms.

Arthropods use hemolymph to circulate oxygen and remove excess carbon dioxide in an open circulatory system. The Arthropod adaptations are the spiracles that take in oxygen along their bodies by being physical holes in their exoskeletons. Some Arthropods have book lungs that function in a rudimentary way. Hemolymph, if it has hemoglobin at all, uses iron as a binding agent. The horseshoe crab is an exception that uses a copper based gas exchange where they have blue colored blood. However: most animals use hemoglobin pigments for gaseous exchange unlike the Arthropods which may or may not use hemoglobin at all.

The principles are that the molecule heme has the capability to bind oxygen in hemoglobin and to some extend hemoglobin binds carbon dioxide although the primary way carbon dioxide is bound is through the bicarbonate ion in solution. The carbon ring structure of heme has ferrous iron in its center that binds the oxygen.

What the arthropods do is pump oxygenated hemolymph straight out through arteries directly into body cavities called sinuses and the return is facilitated internally through pore like structures whereas there are no veins to speak of.

The Arthropods that do not possess hemoglobin intake oxygen directly through their mouth openings and spiracles along the sides of their bodies utilizing passive air flows like ventilation systems.

Due to the gas exchange limitations Arthropods have been restricted in their body plans for millions of years due to the present gas ratio of the atmosphere. At one point in history the gas ratio of the atmosphere was more conducive to the Arthropod body plan, and gas exchange.

The Arthopods use chitin as their primary structure protein and tends to not fossilize in the ways that we are used to as compared to the ossification and fossilization process of animal bone structure. Thusly the Arthropods are the stealth animals of ancient epoch history as their presence has been covered up in time with very little indication of their presence. If there was upright hominid Arthropods exhibiting intelligence there has been no indication that they existed, perhaps using the copper scheme of the Horshoe crab and the book style lungs of the Arachnids.

To Summarize:

The Arthropods primarily respire through pours in their exoskeletons where hemolymph circulates through an open circulatory system from the heart in to open sinuses thereby apparently limiting the maximum size of their phenotype body plan. To further minimize their body plans it appears that few insects utilize hemoglobin, or red blood cells currently but may have done so in past life forms.

Building a hypothetical hominid Arthropod: they would have book type lungs, copper pigments, and perhaps a closed circulatory system in that the last item we have not seen as yet in examples of the Arthropods. There could be a way to combine all of the elements of the Arthropods and come up with a blueprint of the ideal large hominid Arthropod.

Figure 19. In this photo: the giant coconut crab in modern China on the island nation of Taiwan. The coconut crab can reach the size of a dog and has a habitat range nearly spanning the entire Pacific Rim. (Photo credit: Taiwan Today 02022014).

What is intriguing of the Arthropods is that at least one insect uses the spiral arms of the Milky Way galaxy to navigate:

Dung beetles use stars to navigate - study

INDEPENDENT NEWSPAPERS

January 26 2013 at 09:26am

By John Yeld

"Professor Marcus Byrne displays a dung beetle in the Johannesburg Planetarium at Wits University. The star projector was used to determine whether the beetles navigated via the stars of the Milky Way galaxy. Picture: Chris Collingridge..."

"…a Wits University scientist and his research partners have just published their discovery that these fascinating little creatures can navigate using stars… or, more correctly, groups of stars, as they undertake their Sisyphean task of transporting balls of dung…"

"The experiments show that despite their tiny brains and minimal computing power, the beetles can use the relatively dim light of the Milky Way (dim compared to the sun and moon) for orientation – the first species proven to have this ability."

Ichthyology

Figure 20. Fossil Jawed Fish

Fish biology has been going on a long time in scientific research and most of the controversies have been ironed out. In fact a few of the novel forms have been found alive and studied. I have seen a preserved specimen of the Coelacanth, a tetrapod ancestor from the Devonian, in the preserved state once thought extinct. The Coelacanth is our direct ancestor due to the elongated limbs that they developed for purposes unknown deep in the Devonian. However modern specimens of the Coelacanth appear very much alive in at least two locations on Earth. The controversies raging in fish are the phylogenetic tree when examined with the DNA analysis.

An event known as Genome doubling occurred at least once perhaps more for fish. The Perciformes or Acanthopteri show the evidence by the two dorsal fins: one for spines and one of rays among other things. Wild forms occur once the event happens.

Take the swim bladder developing in to the lungs – amazing. There are at least two types of swim bladders. What begs the question is: did the first Tetrapods exploiting land have lungs or gills or what exactly?

Tetrapods Exploiting Land

The Tetrapod fish arriving on land can be tracked with several species of fish taking forays on to land. Once they established themselves it looks like early species remained in bogs and moist areas for quite some time.

Figure 21. In this image: A recent discovery- the Tiktaalik. The image was not available during the previous press releases and represents current work in progress.

The Tiktaalik

Few scientific discoveries are extraordinary events. Even fewer still are events that are truly groundbreaking events (no pun intended). The discovery of the Tiktaalik animal fossil in a Greenland stratum describes a species so transitory that half of its body is one form and the other half of its body is another form altogether - there lies the problem. It is a fantastical finding in the search for how life began. I don't believe the entire story of the Tiktaalik discovery as the article I read (Shubin, 2006).

The article shows a skull line-up of several late Devonian fossils where the fossil of the Tiktaalik in that case looks like a transitory creature per se. Acanthiostega, Epistostege, and of course the Tiktaalik are shown side by side where it starts to look good for the theory of the new species.

As the physiology of the creature is unknown we are lead down a path to accept that somehow the newly discovered creature was half in and out of the water so much that part of its body was still half fish. It must have sat there for long periods of time perhaps posing for its picture in Nature magazine - half in and half out of the water with the water line exactly at the midpoint of its body. The lower half of the body at the posterior end is still half fish.

Everyone's mind has been racing about it. Is this an example of a fish, an amphibian, or a reptile? The first obvious thing that stands out is; it is a creature that somehow is in the process of going from aquatic environments to land. Three things must have happened and there are three examples: salt water to fresh water, water breathing to air breathing, and surviving arid conditions. What we may have is that the Devonian fossil may turn out to be a precursor to something besides fish evolution.

There were three specimens found of the Tiktaalik, and one became the holotype. The fossil isn't complete of the Tiktaalik in the holotype. The drawings are half complete and even half of the body consists of dotted lines for the supposed tail section of the Tiktaalik. In all literature describing the Tiktaalik the tail section is not complete.

One drawing of holotype fossil NUFV 108 is labeled an interpretive drawing. The drawing is left blank for the tail section, and in fact appears deceptive as to the hind limbs of the creature. There is however two views of the creature both dorsal and lateral view, each which is left blank as to the tail section of the holotype fossil.

Figure 22. In this drawing: Tiktaalik, Acanthostega, or Pandericthys.

    The scheme of a life going from water to land is established prior to the Tiktaalik with other species. There are even good fossils and named species already in science for the transition from waters to land. Acanthostega is already a remarkable transitory animal that has a full description. Pandericthys is clearly a fish that is also described. The Tiktaalik supposedly had the front limbs of a land animal (Acanthostega) and the hind limbs of a fish (Panderichthys). It's as if to say - look at this in slow motion - it truly is a transitory animal: the Tiktaalik in the article is half of one species and the other half is another species.

The fossil of the Tiktaalik is a remarkable discovery in the late Devonian strata. It certainly is a tetrapod like the author has described unless there were a set of limbs behind the missing sections of the fossil making it a hexapod. The fossil shows the characteristics of late Devonian specimens of the fresh water aquatic types. The skull preps look like the Tiktaalik could be a new species of transitory animal that fall in the middle of already established clear transitory species, however: the hind quarters of the animal are missing even in the holotype, despite three fossils that were found. It clearly is a remarkable animal to such an extent that there are dotted lines in all descriptions of the animal for the posterior end of the creature, even in the holotype.

The adaption of the jaw from gill arches was a huge evolutionary development for fish. Clearly something went on for fish to exploit land after the jaw but it is not clear what happened. It is not possible to argue very well against the Ichthyology discoveries as given in the example. It is dialed in remarkably well with the work in progress transitory fish discovery. Even the most critical position of the Devonian fish discoveries cannot diminish the move that these tetrapod's made to go on land.

The Jurassic

Figure 23. In this drawing: An unknown anthropomorphic amphibian ancestor.

Once on land there arose amphibians and reptiles. The age of reptiles generally is considered to start from the Jurassic, Triassic and to the Cretaceous: 251-146 mya. Of the period there are only a few specimens, and a lot of theories as to what had gone on. In the anthropomorphic series we could have an unknown amphibian or reptilian hominid still waiting to be found. The ancient lineage still has surviving members today such as the crocodiles, and alligators whom still look much as the Tiktaalik did from the Devonian.

    For fossilization to occur there must be mineralization of the bone where the specimen turns to rock. The bogs and environments of the period do not lend themselves for making fossils very well and it is amazing that there are any fossils at all.

The Mammals

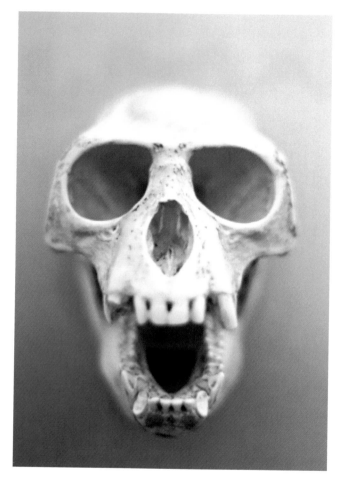

Figure 24. Primate Skull

Mammals arose. The land mammals still retain the eyesight of seeing those things that are underwater. The wavelengths of light that are visible to mammals are as if the organisms are underwater a trait that harkens back to the evolution of fish. The mammals all developed senses independently in a massive adaptation in classical Darwinian forms demonstrating that life on other planets could develop independently to resemble organisms on Earth. Although the animals diversified each animal has developed sensory traits finely tuned for their environment yet work in similar ways.

It is possible to trace the adaptations from basic forms to the mammals in terms of higher development by the adaptations of the heart.

The Comparative Anatomy of the Animal Heart

There is an analysis that can be performed on the circulatory systems of all animal life on planet Earth. The analysis gives an insight as to what has been going on in regards to selection pressures, and adaptations.

The insect heart consists of one chamber that pumps hemolymph into body cavities and sinuses in one direction. The system is an open system. Insects also have very little amounts of red blood cells. The fluid squishes around the cavities and finds its' way to the pump again.

The heart of a fish from the Devonian, such as a shark, is closed and contains only 2 chambers.

The only really muscled chamber is the one ventricle that pumps the blood around, including through the gills, to the body then back to the sinus venosus, and then to the one atrium to be squirted by the ventricle.

The Jurassic reptile heart such as for a crocodile has three chambers. There are some questions as to whether the 3 chambered systems evolved later: however, the modern forms have three chambers including a place where oxygenated and deoxygenated blood mixes in one place. The Jurassic reptile heart is a clue as to the three species of fish like creatures that were attempting to go on to land.

The mammalian heart of a human has four chambers with a tube going in and out of the lungs, and a main tube exiting the heart with oxygenated blood. Essentially all four chambers are heavily muscled where we have a pump going into the lungs, a pump going out of the lungs, a pump going out to the body and a pump coming in from the body.

There could be industry in the comparative anatomy of the animal heart. There could be chapters with multicolored pictures of the dissections. There could be sections on the hematology of the animals, etc. There would be studies on the oxygenation and deoxygenating of blood that would lead to respiration, etc.

The proceedings of the conversation are in the software updates relating to mammals themselves.

We could infer that since the heart seems to have been improved upon then other structures could be expected to be observed as more advanced with respect to our direction of conversation.

For the purposes of the discussion I will talk about the way language and speech work. Ever wonder if there was a gene found for speech what would happen if we just cut and pasted it in to a mouse like software code and watched what happened?

A Human Language Gene Changes the Sound of Mouse Squeaks

New York Times

By NICHOLAS WADE

Published: May 28, 2009

"People have a deep desire to communicate with animals, as is evident from the way they converse with their dogs, enjoy myths about talking animals or devote lifetimes to teaching chimpanzees how to speak. A delicate, if tiny, step has now been taken toward the real thing: the creation of a mouse with a human gene for language."

"The gene, FOXP2, was identified in 1998 as the cause of a subtle speech defect in a large London family, half of whose members have difficulties with articulation

and grammar. All those affected inherited a disrupted version of the gene from one parent. FOXP2 quickly attracted the attention of evolutionary biologists because other animals also possess the gene, and the human version differs significantly in its DNA sequence from those of mice and chimpanzees, just as might be expected for a gene sculpted by natural selection to play an important role in language."

"Researchers at the Max Planck Institute for Evolutionary Anthropology in Leipzig, Germany, have now genetically engineered a strain of mice whose FOXP2 gene has been swapped out for the human version. Svante Paabo, in whose laboratory the mouse was engineered, promised several years ago that when the project was completed, "We will speak to the mouse." He did not promise that the mouse would say anything in reply, doubtless because a great many genes must have undergone evolutionary change to endow people with the faculty of language, and the new mouse was gaining only one of them. So it is perhaps surprising that possession of the human version of FOXP2 does in fact change the sounds that mice use to communicate with other mice, as well as other aspects of brain function."

"That is the result reported in the current issue of the journal Cell by Wolfgang Enard, also of the Leipzig institute, and a large team of German researchers who studied 300 features of the humanized mice. FOXP2, a gene whose protein product switches on other genes, is important during the embryo's development and plays an active part in constructing many tissues, including the lungs, stomach and brain. The gene is so vital that mice in which both copies of the gene are disrupted die after a few weeks."

"Despite the mammalian body's dependence on having its two FOXP2 genes work just right, Dr. Enard's team found that the human version of FOXP2 seemed to substitute perfectly for the mouse version in all the mouse's tissues except for the brain."

"In a region of the brain called the basal ganglia, known in people to be involved in language, the humanized mice grew nerve cells that had a more complex structure. Baby mice utter ultrasonic whistles when removed from their mothers. The humanized baby mice, when isolated, made whistles that had a slightly lower pitch, among other differences, Dr. Enard says. Dr. Enard argues that putting significant human genes into mice is the

only feasible way of exploring the essential differences between people and chimps, our closest living relatives."

"There are about 20 million DNA differences between the genomes of humans and chimps, but most make no physical difference. To understand which DNA changes are important, the genes must be put into another species. There is no good way of genetically engineering chimps, even it were ethically acceptable, so the mouse is the test of choice, in Dr. Enard's view…"

"Dr. Joseph Buxbaum, an expert on the molecular basis of psychiatric disease at Mount Sinai Medical Center, said Dr. Enard's team had taken a good first step toward understanding the role of FOXP2 in the development of the brain. "The most surprising finding, and cause for great optimism, is that the gene does seem to have a great effect on pathways of neural development in mice," he said."

"Dr. Gary Marcus, who studies language acquisition at New York University, said the study showed lots of small effects from the human FOXP2, which fit with the view that FOXP2 plays a vital role in language, probably with many other genes that remain to be discovered. "People shouldn't think of this as the one language gene but *as part of a broader cascade of genes," he said. "It would have been truly spectacular if they had wound up with a talking mouse."…"*

Mammals arose into primates. There was a common ancestor between apes and humans.

Fossil of Great Ape Sheds Light On Evolution

Science News (Science Daily)

May 1st 2013

"…Researchers who unearthed the fossil specimen of an ape skeleton in Spain in 2002 assigned it a new genus and species, Pierolapithecus catalaunicus. They estimated that the ape lived about 11.9 million years ago, arguing that it could be the last common ancestor of modern great apes: chimpanzees, orangutans, bonobos, gorillas and humans. Now, a University of Missouri integrative anatomy expert says the shape of the specimen's pelvis indicates that it lived near the beginning of the great ape evolution, after the lesser apes had started to develop separately but before the great ape species began to diversify."

A lot of work has been done on the Chimpanzee, and the Bonobos, where a separation line exists along the Congo River. North of the river exist the Chimpanzee while South exist the Bonobos. The species look identical but at the genetic level there are differences. There are behavioral differences as well.

The behavioral differences are fascinating where the Northern group exist alongside the Gorillas and exhibit violent behavior while the Southern group exist in a world of plentiful food and no competing primates and have a peaceful existence, etc. The behaviors are described by animal-behavior-zoologists and increasingly anthropologists due to the primates using tools such as blades of grass for termite fishing, etc.

You could break off and write paper after paper on geographic separation by the river boundary and speciation. You could branch out and write a paper on the beginning of warfare. You could make a lifetime career of the sexual activity of the Bonobos, the homosexual activity of the Bonobos and the group sex of the Bonobos and make films, and books, etc.

Despite famous experiments none of the animals have been able to vocalize speech the way a human does, and splitting the hairs as such really do not have the capacity that we do for language use. Some of the experiments can teach apes sign language after decades of wrote memory training however: no one has presented an argument that the apes are really intelligent beings. It is ubiquitous and any one can clearly see that Chimpanzees and Bonobos although share a good percentage of DNA markers with us they are worlds apart.

Chimps have the speech gene but it is not like the human version:

Speech Gene Shows Its Bossy Nature

New York Times

By NICHOLAS WADE

Published: November 11, 2009

"Of the 20,000 genes in the human genome, few are more fascinating than FOXP2, a gene that underlies the faculty of human speech."

"All animals have an FOXP2 gene, but the human version's product differs at just 2 of its 740 units from that of chimpanzees, suggesting that this tiny evolutionary fix may hold the key to why people can speak and chimps cannot."

"A scientific team led by Dr. Daniel H. Geschwind of the University of California, Los Angeles, has now completed a parallel experiment, which is to put the chimp version of FOXP2 into human neurons and see what happens. These were neurons living in laboratory glassware, not a human brain, so they gave a snapshot of FOXP2 only at the cellular level. But they confirmed suspicions that FOXP2 was a maestro of the genome.

The gene does not do a single thing but rather controls the activity of at least 116 other genes, Dr. Geschwind's team says in the Thursday issue of Nature."

"Like the conductor of an orchestra, the gene quiets the activity of some and summons a crescendo from others. Surprisingly, the chimp version of the gene had a more forceful effect in the human nerve cells than did the human version."

"…"The human FOXP2 seems to be acting on a more refined set of genes," Dr. Geschwind said in an interview from London."

"Several of the genes under FOXP2's thumb show signs of having faced recent evolutionary pressure, meaning they were favored by natural selection. This suggests that the whole network of genes has evolved together in making language and speech a human faculty."

"…And some of the genes in FOXP2's network have already been implicated in diseases that include disorders of speech, confirming its importance in these faculties."

One mammalian primate approximately 200,000 years ago developed speech. The speech development of primate mammals can be traced using microbiological techniques.

The Hominids

Figure 25. H. neanderthalensis

One of the most difficult problems and unsolved mysteries, on which a great number of scientists have been working for many decades, is a human origin. Nevertheless, there is a theory, which is based on Charles Darwin's teaching called "evolutionary theory": the origin of a human from a common ancestor to apes through transitional forms as a result of natural selection. Thus, a new research has been conducted in this area in order to form the current hominid list and simplify the meaning of the theory.

'Three million years ago, there were primates which we now call australopithecines. The australopithecines had brains no bigger than chimpanzee's, they were shorter than us, with very long arms, they were covered in fur, but they were bipedal, as we are' (Munford). There are three species of australopithecines: Australopithecus africanus, Australopithecus robustus, and Australopithecus boisei. All of them had their own features, for instance, scientists from Australian museum are convinced that: "Australopithecus africanus were the first of our pre-human ancestors to be discovered, but was initially rejected from our family tree because of its small brain. This opinion changed when new evidence showed this species had many features intermediate between apes and humans" (Dorey, 2011).

"Around 2 million years ago, a significant change occurred in the australopithecines' brain size. The change in overall structure of the species meant that taxonomists gave an entirely new genus to the species. The species of homo pertains to the more recent ancestral line of modern man, homo meaning the same, and sapiens pertaining to 'man'" (Biology online, 2001).

    Homo erectus (sometimes called Homo ergaster) are the oldest known early humans to have possessed modern human-like body proportions with relatively elongated legs and shorter arms compared to the size of the torso. These features are considered adaptations to a life lived on the ground, indicating the loss of earlier tree-climbing adaptations, with the ability to walk and possibly run long distances. Compared with earlier fossil humans, note the expanded braincase relative to the size of the face. The appearance of Homo erectus in the fossil record is often associated with the earliest hand axes, the first major innovation in stone tool technology (The Smithsonian's National Museum of Natural History, 2011).

    The species that living human beings on this planet belong to is Homo sapiens. During a time of dramatic climate change 200,000 years ago, Homo sapiens (modern humans) evolved in Africa. Like other early humans that were living at this time, they gathered and hunted food, and evolved behaviors that helped them respond to the challenges of survival in unstable environments. Scientists sometimes use the term "anatomically modern Homo sapiens" to refer to members of our own species who lived during prehistoric times (The Smithsonian's National Museum of Natural History, 2011).

As a result, Homo sapiens transformed into Homo sapiens Sapiens. Canalon (2001) summarizes: from 30 000 years ago up until this present day, our own species has exhibited the most advantageous characteristics to adapt and manipulate our environment. From this point, the species and its component skills managed to colonize all the main continents of today's world, bar Antarctica, which still presented conditions unbearable to the species and the technology of the time. However, more complex tools were being developed, and that has continued over the period of time where we have successfully monitored historical events in our human race. At this point, human history in the abstract manner truly begins.

Language in Hominids

The next thing I would like to discuss is how animals on this planet obtained the gift of language. At this point we know of no other creature that currently has the language processor that we have of the universal language processor hardwired in our brains.

Scientists Report Finding a Gene for Speech

New York Times

HEALTH

By NICHOLAS WADE

Published: October 4, 2001

> "A team of geneticists and linguists say they have found a gene that underlies speech and language, the first to be linked to this uniquely human faculty."

> "The discovery buttresses the idea that language is acquired and generated by specific neural circuitry in the brain, rather than by general brain faculties."

In the primary literature, Science Magazine in 2002:

> "Last year researchers identified the first gene implicated in the ability to speak, FOXP2. This week, a research group shows that the human version of this so-called speech gene appears to date back no more than 200,000 years--about the time that anatomically modern humans emerged. The authors argue that their findings are consistent with speculation that

the worldwide expansion of modern humans was driven by the emergence of language abilities (Balter, 2002)."

So there we have it – the universal processor of unknown origin and purpose of language. The gene goes far in to antiquity and other methods must be used to determine the origins of language in the genetic code.

In regards to brain density, the left hemisphere appears to have a difference in arrangement as compared to the right hemisphere and is made available for the behaviors necessary for speech.

For the theory of Panspermia to be considered: I think it is interesting that the gene for language just pops in. I don't think that if we ate an ET that we would get the gene. A woman would have to have been crossed with an ET and perhaps a really, really bizarre mutant were close enough to our common ancestor that the offspring received the gene for language.

H. neanderthalensis was supposedly able to process language. I have heard speech synthesizers that were developed to determine what the sounds were that they made on their palates. It sounded like middle Egyptian to me where: when the glyphs are spoken you get the spoken part of the sounds of the language. Without a reconstituted H. neanderthalensis here I am not able to psychoanalyze the H. neanderthalis so we don't know right now if they spoke like we do.

The gene for language appears to have been handed down through the ages at least in the Homo genus to the point of Homo neanderthalensis. A recent New York Times article describes the finding:

Neanderthals Had Important Speech Gene, DNA Evidence Shows

New York Times

Oct 18, 2007

"The genes of Neanderthals seemed to have passed into oblivion when they vanished from their last refuges in Spain and Portugal some 30,000 years ago, almost certainly driven to extinction by modern humans. But recent work by Svante Paabo, a biologist at the Max Planck Institute for Evolutionary Anthropology in Leipzig, Germany, has made it clear that some Neanderthal DNA can be extracted from fossils…"

"Dr. Paabo, Dr. Johannes Krause and Spanish colleagues who excavated the new bones say they have now extracted the Neanderthal version of the relevant part of the FOXP2 gene. It is the same as the human version, they report in today's issue of Current Biology…"

"Because many other genes are also involved in the faculty of speech, the new finding suggests but does not prove that Neanderthals had human-like language."

"There is no reason to think Neanderthals couldn't speak like humans with respect to FOXP2, but obviously there are many other genes involved in language and speech," Dr. Paabo said."

"The human version of the FOXP2 gene apparently swept through the human population before the Neanderthal and modern human lineages split apart some 350,000 years ago (Wade,2007)."

Evidence exists for a hybrid:

Skeletal Remains of Neanderthal/Human Hybrid Found

I09.com

March 28 2013

"The skeletal remains of an individual living in northern Italy about 40,000 to 30,000 years ago are believed to be that of a love child produced by a human-Neanderthal couple, according to a paper in PLoS ONE. If further analysis proves the theory correct, the remains belonged to the first known such hybrid, providing direct evidence that humans and Neanderthals interbred."

First Love Child of Human, Neanderthal Found

Discovery News

March 27, 2013

"The skeletal remains of an individual living in northern Italy 40,000-30,000 years ago are believed to be that of a human/Neanderthal hybrid, according to a paper in..."

"If further analysis proves the theory correct, the remains belonged to the first known such hybrid, providing direct evidence that humans and Neanderthals interbred. Prior genetic research determined the DNA of people with European and Asian ancestry is 1 to 4 percent Neanderthal."

"The present study focuses on the individual's jaw, which was unearthed at a rock-shelter called Riparo di Mezzena in the Monti Lessini region of Italy. Both Neanderthals and modern humans inhabited Europe at the time."

Scientists have been attempting to put together what a Neanderthal would sound like from their palates.

Computer 'Recreates' Neanderthal Speech

Fox News

April 17, 2008

"Robert McCarthy, an assistant professor of anthropology at Florida Atlantic University in Boca Raton, Fla., used ancient skeletons to reconstruct an approximation of the Neanderthal vocal tract — and then had a computer recreate the sounds it would make."

I have heard the sounds that the simulators make. The demonstration to me sounds like Middle Egyptian, which presents another can of worms.

In principle to summarize:

What we see is that something in the form of a model was passed on across the species in the genus Homo. This model could be what we term mind and the way it is conducted. What is even more profound is the fact that the coding for the model is done in a genetic manner which would indicate that there exists a universal software code for intelligence, mind and for speech processing. There must be an abstract model for mind and speech *somewhere that can be distilled out and represented as a symbol or series of symbols in principles of axioms.*

In conclusion:

In the calculation of interaction of particles in the Universe, there simply are not enough particles in the universe to interact randomly to create the first animal cell in one location. However: Animal life may have begun as single celled organisms here on the planet formed as the result of arrival of foreign genetic codes. The inference of Panspermia may be postulated given the evidence that foreign genes may have arrived on planet Earth for the basal organisms to form. Foreign genes can be incorporated into cells on planet Earth as shown in the experiment. It may be the case that Earth has seeded the planet Mars by the case of apparent myxomycetes observed there. At some point a series of developments was made for life on Earth. Life on the planet arose into animals, that formed into fish and the fish eventually exploited the resources on land. The new land animals eventually became primates. The primates arrived at an adaptation of language in a new genus Homo. This can be inferred that our genes for language, and our brain structures and learned behaviors are native to the genus Homo. The language adaptation resulted in a type of universal language processor of the current human brain. Even though individual human brains are different, and language is learned, the operations of the brain contribute to the similarity of language because Brain structures indicate language is hardwired and genes have been found for language in H. sapiens and H. neanderthalensis.

Notes:

The research end notes. My perspective has been achieved by reviewing the following articles. The fun clippings that have importance that have alternately been included and cut out, put in margins, and footnotes, found on bar napkins, etc.:

Stuff of Life (but Not Life Itself) Is Detected on a Distant Planet

New York Times

Published: March 20, 2008

By DENNIS OVERBYE

"Astronomers reported Wednesday that they had made the first detection of an organic molecule, methane, in the atmosphere of a planet outside our solar system and had confirmed the presence of water there, clearing the way for a bright future of inspecting the galaxy for livable planets, for the chemical stuff of life, or even for life itself…"

"Under the right conditions, water can combine with organic chemicals like methane to make amino acids, the building blocks of life as we know it. While the presence of these chemicals was not a big surprise and while the planet in question — in the constellation Vulpecula — is too hot *and massive for living creatures, the result left astronomers elated at their improving powers of celestial discernment…"*

"The Vulpecula planet was in the news a year ago after astronomers using the Spitzer Space Telescope tried and failed to find signs of water there and on another planet — a surprise since all the theoretical models predicted it should be there in abundance. But that was a year ago, a long time in the world of exoplanets, where an avalanche of data in the last decade has produced a series of milestones and some 270 new planets…"

Organic Compound Found In the Stars

Life-building molecules might be spread throughout space.

Nature

Published online: 24 July 2007

Ewen Callaway

"Astronomers have found the largest negatively charged molecule so far seen in interstellar space. The discovery, of an organic compound, suggests that the chemical building blocks of life may be more common in the Universe than had been previously thought…"

"The molecule is a chain of eight carbons and a single hydrogen called the octatetraynyl anion (C_8H^-). Two teams of scientists have spotted it near a dying star and in a cloud of cold gas…"

"The discovery, along with that of three smaller organic molecules in the past year, opens up a suite of potential chemical reactions and products. It suggests that 'prebiotic' molecules such as amino acids, the building blocks of protein, could form all over the Universe, says Tony Remijan, an astronomer at the National Radio Astronomy Observatory (NRAO) in Charlottesville, Virginia…"

"Remijan's team spotted the octatetraynyl anion in a dense cloud of gas in the halo surrounding a dying star in the constellation Leo, 550 light years from Earth. They made the discovery using the Green Bank Telescope in West Virginia, a radio dish 100 meters across…"

"Another team from the Harvard-Smithsonian Center for Astrophysics in Cambridge, Massachusetts, used the same telescope to spot the compound in the Taurus Molecular Cloud (TMC), located 450 light years from Earth in the constellation Taurus. Both findings are published in the Astrophysical Journal…"

"…negatively charged organic molecules such as the octatetraynyl anion were long thought to be confined to Earth. "People thought they were too fragile to exist [in space]," says Sandra Brünken, a member of the Harvard-Smithsonian team…"

"To show that octatetraynyl anion could be found in space, Brünken made the compound, along with several similar ones, in the lab. She then measured the chemical's spectrum — the same property radio telescopes hunt for. With this in hand, the two teams sifted through radio telescope measurements to find the molecule…"

"Brünken and her colleagues discovered the first interstellar organic anion — a chain of six carbons and one hydrogen — by chance in late 2006 when they noticed a blip in the measurements from the TMC. The spectrum hinted at a compound called hexatriynyl anion, but confirmation only came when they made the compound and lab and found that the two signatures matched up…"

"The researchers believe that even larger organic molecules are likely to be out there. Although more complex molecules are harder to identify, Remijan is confident of further discoveries. "They're really easy to find once you know what you're looking for," he says…"

First Sign of Water Found on an Alien World

NewScientist.com news service

Updated 22:05 11 April 2007

David Shiga

"Water has been detected in the atmosphere of an alien world for the first time, a new analysis of Hubble Space Telescope data suggests…"

"The planet, called HD 209458b, is about 70% as massive as Jupiter and is scorched by the heat of its parent star, which it orbits 9 times as close as Mercury does to the Sun…"

"He says the relatively small amount of light filtering through at about 0.9 microns suggests the presence of water, which absorbs light at this wavelength. "To me, that's a clear indication that water is there," Barman told New Scientist. "I think this is the first time we've had strong evidence that there's water in at least one extra solar planet. "But despite the presence of water, he points out that the planet's prevailing temperatures of about 1000° Celsius mean conditions would not be favorable to life. "It's not a place you or I would want to visit," he says…"

Amino Acid Found In Deep Space

New Scientist

18 July 2002

Rachel Nowak

"An amino acid, one of the building blocks of life, has been spotted in deep space. If the find stands up to scrutiny, it means that the sorts of chemistry needed to create life are not unique to Earth verifying one of astrobiology's cherished theories. This would add weight to ideas that life exists on other planets, and even that molecules from outer space kick-started life on Earth. Over 130 molecules have been identified in interstellar space so far, including sugars and ethanol. But amino acids are a particularly important find because they link up to form proteins, the molecules that run, and to a large extent make up our cells. Back in 1994, a team led by astronomer Lewis Snyder of the University of Illinois at Urbana-Champaign announced preliminary evidence of the simplest type of amino acid, glycine, but the finding did not stand up to closer examination (New Scientist magazine, 11 June 1994, p 4).Now Snyder and Yi-Jehng Kuan of the National Taiwan Normal University say they really have found glycine. "We're more confident [this time]," says Kuan. "We have strong evidence that glycine exists in interstellar space. " Huge Blobs. The researchers monitored radio waves for the spectral lines characteristic of glycine. They studied emissions from more locations than before – giant molecular clouds, huge blobs of gas and dust grains. They have also identified 10 spectral lines at each location that correspond to the lines created by glycine in the lab; before they had just two. The discovery of glycine supports recent lab-based simulations of deep space, which show that ices containing simple organic matter could form. When researchers bathe those ices in ultraviolet light, amino acids are created. "Glycine is the holy grail," says Jill Tarter, director of the Centre for SETI Research at the SETI Institute in Mountain View. "Let's hope they've got it this time." The new research was presented at the Bioastronomy 2002 meeting, held on Hamilton Island, Queensland between 8 and 12 July. (http://www.newscientist.com/article.ns?id=dn2558)."

The Thunderwell Story

"The February/March 1992 issue of Air & Space magazine, published by the Smithsonian, contained an article about nuclear rocket propulsion:"

Air & Space

February/March 1992

Overachiever

"Every kid who has put a firecracker under a tin can understands the principle of using high explosives to loft an object into space. What was novel to scientists at Los Alamos [the atomic laboratory in New Mexico] was the idea of using an atomic bomb as propellant. That strategy was the serendipitous result of an experiment that had gone somewhat awry. "Project Thunderwell was the inspiration of astrophysicist Bob Brownlee, who in the summer of 1957 was faced with the problem of containing underground an explosion, expected to be equivalent to a few hundred tons of dynamite. Brownlee put the bomb at the bottom of a 500-foot vertical tunnel in the Nevada desert, sealing the opening with a four-inch thick steel plate weighing several hundred pounds. He *knew the lid would be blown off; he didn't know exactly how fast. High-speed cameras caught the giant manhole cover as it began its unscheduled flight into history. Based upon his calculations and the evidence from the cameras, Brownlee estimated that the steel plate was traveling at a velocity six times that needed to escape Earth's gravity when it soared into the flawless blue Nevada sky. 'We never found it. It was gone,' Brownlee says, a touch of awe in his voice almost 35 years later. "The following October the Soviet Union launched Sputnik, billed as the first man-made object in Earth orbit. Brownlee has never publicly challenged the Soviet's claim. But he has his doubts."*

Bibliographie

Chapoter style in accordance witz The Associates Pressa Stylebook.
Références listed in reverse chronologique ordre in APA format.

Brandon J. (2012) Was There a Natural Nuclear Blast on Mars? Retrieved from http://www.foxnews.com/scitech/2011/04/01/natural-nuclear-blast-mars/#ixzz27WEl4uTD

Dorey , F. (2011). Australopithecus africanus. Australian museum Munford, M. (n.d.). Evolution and the prehistory of man. Retrieved from http://evolution-of-man.info/index.htm

The Smithsonian's National Museum of Natural History (2011). What does it mean to be a human. Retrieved from http://humanorigins.si.edu/evidence/human-fossils/species/homo-erectus

Wade, N. (2007, October 18). Neanderthals May Have Had Gene for Speech. New York Times.

Neil H. Shubin1, Edward B. Daeschler2 & Farish A. Jenkins Jr3. The pectoral fin of Tiktaalik roseae and the origin of the tetrapod limb. Nature. Vol 440|6 April 2006.

Edward B. Daeschler1, Neil H. Shubin2 & Farish A. Jenkins Jr3. A Devonian tetrapod-like fish and the evolution of the tetrapod body plan. Nature, Vol 440|6 April 2006.

Michael Balter. (2002). 'Speech Gene' tied to modern humans. Science, 297(5584), 1105.

Canalon. (2001). Homo sapiens sapiens. Retrieved from http://www.biology-online.org/10/15_homo.htm

Biology online. (2001). Early species of homo. Retrieved from http://www.biology-online.org/10/14_early_hominids.htm

Labandeira, C., & Sepkoski, J. (1993). Insect diversity in the fossil record. Science, 261(5119), 310-315.

Hoyle, F. (1981). Evolution from Space. Dent, The University of California.

Proctor, Richard A. (1882, Jan 20) Knowledge: Meteoric Organisms. Wyman & Sons. 74-76, Great Queen Street, Lincoln's-Inn Fields, W.C.

Chapter 3

Panspermia: Hypothesis

My Panspermia Hypothesis

Even though no other animal life has been observed off world, Animal life began from an input from at least one other source in the Universe because the odds of particle collisions are too great for one input source for animal creation and Amino acids are observed in spaces which are the building blocks of life.

Amino acids are observed in space which is the building blocks of life:

Amino Acid Found In Deep Space

New Scientist

18 July 2002

Rachel Nowak

> "An amino acid, one of the building blocks of life, has been spotted in deep space. If the find stands up to scrutiny, it means that the sorts of chemistry needed to create life are not unique to Earth verifying one of astrobiology's cherished theories."

The most important reason Animal life began from an input from at least one other source in the Universe is because the odds of particle collisions are too great for one input source for animal creation.

> According to Hoyle the chances for a cell to acquire the set of enzymes required for life would be $10^{40,000}$ while the number of molecules in the Universe is 10^{80} thereby proving in his calculations that something unusual went on in the primordial creation of the first cell on Earth (Hoyle, P. 35, 1981).

In conclusion, although No other animal life has been observed off world, Animal life began from an input from at least one other source in the Universe for two main reasons. First, Amino acids are observed in space which is the building blocks of life. But most importantly, the odds of particle collisions are too great for one input source for animal creation.

Chapter 4

Conclusion

To conclude, it is necessary to mention that this dissertation deals with one of the most interesting and controversial theories of the appearance of life on our planet – the theory of Panspermia. There are a lot of disputes around the life origin, and many different theorists propose a lot of different theories concerning this issue. However, it cannot be argued that the theory of Panspermia is one of the most comprehensive ones and based on the strong ground of rational arguments proved by the practical research. If the theory will be proven it would require indirect means perhaps even alternative frameworks of stating the problem even finding a meteorite with fossils and organisms.

Panspermia is the theory which states that the variety of living forms came to the Earth from the cosmos. There are two categories of Panspermia: Direct and Undirected. If to summarize the knowledge received in the primary research in the field of Panspermia theory, the following statement may be formulated. In the calculation of interaction of particles in the Universe, there simply are not enough particles in the universe to interact randomly to create the first animal cell in one location. However, Animal life may have begun as single celled organisms here on the planet formed as the result of arrival of foreign genetic codes. The inference of Panspermia may be postulated given the evidence that foreign genes may have arrived on planet Earth for the basal organisms to form. Foreign genes can be incorporated into cells on planet Earth as shown in the experiment. It may be the case that Earth has seeded the planet Mars by the case of apparent myxomycete observed there. At some point, a series of developments was made for life on Earth. Life on the planet arose into animals that formed into fish and the fish eventually exploited the resources on land. The new land animals eventually became primates. The primates arrived at an adaptation of language in a new genus Homo. This can be inferred that our genes for language and our brain structures and learned behaviors are native to the genus Homo. The language adaptation resulted in a type of universal language processor of the current human brain. Even though individual human brains are different, and language is learned, the operations of the brain contribute to the similarity of language because Brain structures indicate language is hardwired and genes have been found for language in H. sapiens and H. neanderthalensis.

Although, in accordance with the telescopic observations, no other animal life has been observed off world, Animal life began from an input from at least one other source in the universe for two main reasons. First, Amino acids are observed in space which is the building blocks of life. But most importantly, the odds of particle collisions are too great for one input source for animal creation. The above mentioned statement can be confirmed by the results of the following experiment. The purpose of this experiment was to find out whether the theory of Panspermia can be proved being based on the fact that transformation of a bacterial genome can take place when bacteria take up small amounts of DNA from their environment. Escherichia coli were used in order to take up a jellyfish gene that would cause the bacteria to glow green under UV light. In accordance with the results, after inoculating a plate with E. coli the plate was introduced with a gene that causes glow in the dark phenotype. The E. coli took up the gene and incorporated it into their genome. Therefore, it may be stated that the theory of Panspermia was proved on the practice.

Appendix 1

Figure 26. E coli with foreign gene for bioluminescence

Experiment: Bacterial Transformation

Goals

Transformation of a bacteria genome can take place when bacteria take up small amounts of DNA from their environment. The theory of Panspermia can be proven if the experiment is successful.

Materials

Escherichia coli will be used to take up a jellyfish gene that will cause the bacteria to glow green under UV light.

Synopsis

After inoculating a plate with E. coli the plate was introduced with a gene that causes glow in the dark phenotype. The E. coli took up the gene and incorporated it into their genome.

Appendix 2

Experiment:
Is a Meteorite Sample Indicative Of Water Stratification On Mars?

Figure 27. Mars Rover "Opportunity "Picture Mission Day 642

A recent Mars Rover picture of a point of exploration has brought to life many questions on the nature of the water stratification on Mars. Like the picture above - a meteorite long thought to be from the stratification on Mars, would indicate that an Epoch on Mars had long standing water. The current geological time "weathering," appears to be that of wind erosion, however: deeper layers of the Mars strata indicate water sediments indicative of standing water deposits forming mineralization layers[1]. The amounts of geologic time that are needed to form these mineralization layers, would indicate that water stood and flowed however slow, or perhaps torrential on the Mars surface for thousands of years. Hydrous Atmosphere activity found in meteorites previously would indicate this sample, not only is from Mars but is evidence of an Epoch on Mars previously unknown to our exploration. This conclusion is made stronger by the recent photographs from the Mars Rover Opportunity.

Figure 28. Sample Of Meteorite

1. Hydrous Minerals In Meteorites Nature,256:697, 1975 , Hughes, David W "…Meteoritic minerals are nearly always fresh and unrecompensed--- which contrasts with many of the water-containing products of weathering and erosion which occur in such profusion on Earth. …water must have been present during the formation of this shocked olivine, and that the meteoric mineral originated on a body in the solar system that had a hydrous atmosphere."

CPSIA information can be obtained at www.ICGtesting.com
Printed in the USA
LVIW01n2010230417
531914LV00001B/1